ANTENNA SITE OPERATING GUIDE

by John S. Hollar

Published by Hollar Communications
5201 South Torrey Pines Drive
Suite 1255
Las Vegas, Nevada 89118-0608

The **Antenna Site Operating Guide** (hereinafter: *Guide*) in order to be effective, must refer to many suppliers, products, publications, trade journals and celebrated individuals in the field of telecommunication antenna site technology and theory. The author acknowledges that these sources are crucial to this subject. Accordingly, when a specific reference to an actual product name, protocol or article is made, the sources are assumed to be and are identified as being copyrighted or claimed as trademarks. In addition some names describing a process or a definition are in themselves registered names such as a *Thruline* ® wattmeter or a touch-tone™ telephone. Where any of the above designations, quotes or sources appear in the *Guide*, the author has endeavored to identify completely the specific source (usually with a Web site address), and has denoted the information either by marks ® © ™ ℠, *italics* or set off in quotations.

The author and the publisher have taken care in the preparation of this *Guide*, but make no expressed nor implied warranty of any kind and assume no responsibility for errors or omissions. No liability is assumed for incidental or consequential damages in connection with or arising out of the use of this *Guide* or the information or conjectures contained herein. Conclusions presented in this *Guide* are based upon the personal experiences of the author and as such are expressed as opinion, for humor and for discussion purposes only.

Library of Congress Cataloging-in-Publication Data

Hollar Jr., John S.

Antenna Site Operating Guide: A Guide to Building, Marketing and Operating an Antenna Site for Profit/ John S. Hollar Jr.

p. cm.

ISBN 0-9667428-0-X (Pbk: acid-free paper)

1. Antennas 2. Antenna Site Construction 3. Marketing Antenna Sites

This book is available for bulk purchase by companies, institutions, consultants or other organizations; and may be custom *recovered* for promotional or educational purposes. For more information please contact the author.

Find our Web page at:
http://members.aol.com/jrhollar/
e-mail Jrhollar@aol.com

ACKNOWLEDGMENTS

There were many valuable sources who helped put this project together. Unfortunately, there is insufficient space to name them all, but I will single out a few deserving special attention: colleague, Brian E. Danzis, an amazingly creative and compassionate gentleman; courageous friend, Kay L. Mark, legal and real estate guru who assisted me greatly in the completion of this work; my kids, Kirsten and Jason who monitored my work all over the country with gentle encouragement; and finally, Margaret Mary Simok, communications pioneer, manager and friend without whom this book would not have been inspired.

For Repiints or Onginals of RCR's Editorial Cartoon, Contact Chris Boyer at (303) 455-8654

Published with permission from simply the best communication cartoonist of our age. . .*Chris Boyer (303) 455-8654*

CONTENTS

CHAPTER THREE

CHAPTER FOUR

CHAPTER FIVE

CHAPTER SIX

CHAPTER SEVEN

CHAPTER EIGHT

CHAPTER NINE

CHAPTER TEN

Introduction

Communications forecasters tell us that there will be over 80 million cellular telephone subscribers within the next several years. One-way paging (beeper) subscribers could number 60 million during the same period. Everyone can relate to the seemingly endless number of antennas, lighted and unlighted, jutting from the tops of buildings, positioned on towers, and on nearby hill tops. These antennas and their corresponding mounting locations form the backbone of our communications infrastructure. Antenna *sites* provide the permanent transmission and reception locations for a wide range of communication services. In addition to the *traveling public*, i.e. users of cellular telephone and paging services, fixed locations are also required for nearly every wireless service. Public service systems such as fire, police, and emergency medical facilities, public utilities, businesses and even shared community repeater systems also require antenna sites. So much so that thousands of *licensed base stations* can be identified in just about every population center in the country. Every one of these base stations must have a fixed location.

Many more antenna sites are needed

Twenty years ago communication services required only a mountain top or high location for operation. The theory was that the higher the base station was located the greater the resulting coverage. The principle is still true today. However this configuration is generally only able to serve one user at a time. How is it possible then to serve several users on the same frequency at the same time? In order to do this the base station is relocated to a much lower position and the transmitting power reduced substantially so that the effective coverage will no longer be line of sight but just a few miles. By installing many of these low power low profile transmitting and receiving stations many more users can be served on the same frequency from much smaller radiation patterns. The final objective is to have the total number of new lower power stations cover the original radiation pattern developed by the single high profile base station previously located on the hill top. In order to provide service to many users we need a switching technology to direct and coordinate the base station operation as users move from one small coverage area to an adjacent one. Obviously what is being described in simple terms is the basis for cellular radio. These small coverage areas are commonly referred to as radio zones, ring coverages or *cells*.

Because the idea of using low power low profile antenna sites is catching on for almost every radio service, the *number* of antenna site locations will multiply many times. Site selection is no longer governed solely by a convenient location on a tall building, a water tower or a natural hill. Rather, communication providers are now carefully analyzing their coverage areas in relation to the sites they have, the customers they serve or want to serve and the sites they now must obtain to meet their needs.

The author has spent many years in the construction, maintenance, and operation of antenna sites and more recently as a *site-finder* and consultant to communications firms aggressively building-out their initial and expanding operations. It is rewarding work for someone with a bit of training, a love of travel and the ambition to actually find a workable location within a target area. When a location is found an Antenna Site Agreement is negotiated with its owner. These arrangements provide long term monthly income with a minimum of investment. Even more exciting is that once a site has been established it won't be long before additional accounts come on board providing additional income. The reason is simple, once you have an established antenna site, what works for one communication firm often works for another.

Objectives of this *Guide*

It is the objective of this *Guide* to provide the methods needed to evaluate, market and operate an antenna site. You may also be fortunate enough to have control of several locations and wish to develop some or all of them as well. When you are confident that the locations you have selected meet the criteria suggested you may proceed at your own pace. Even with a minimum investment in some cases as low as $500, you can position yourself personally or through an organization, i.e. a church, a business unrelated to telecommunications or even as an agent for somebody else's property, to cash in on a very stable and lucrative enterprise. With a little more experience you may even want to speculate on proposed new sites or assist others in building and marketing antenna locations. There are many avenues of enterprise in this industry as outlined in Chapter 10.

What is in it for you?

A typical remote receiver site might generate as high as $250 to $300 per month from one of your locations. If it works well and the word *really* gets out when it does, additional accounts will soon follow. With as little as 4 square feet of land for an antenna mast and another 64 square feet of land for an equipment housing you could enjoy net incomes in the range of $18,000 per year. Several site locations or expanded single site facilities can earn substantially more. This *Guide* not only describes a small starter system but explores in detail larger and more complex *Antenna Farms*. It explains just what is involved in presenting your location to the hundreds of communication firms who will actively use your site information in their continuing plan for expansion and build out.

The author is confident that there is sufficient material included in this manual to allow anyone with reasonable map reading and calculator skills to prepare the necessary profiles of interest for the telecommunications industry. It is also hoped that the material in this *Guide* will enable many home based businesses or variations thereof, to profit from the development of these much needed antenna sites around the country. In addition there is no doubt that by increasing the "pool" of available and proven sites to the telecommunications industry, faster service deployment and reliability will result for a public anxious to utilize the new technologies which are being developed.

CHAPTER ONE

1.1 Changing Communications Needs

Antenna sites have been around for over seventy-five (75) years. In the early days, AM radio stations required many acres of land to install multi-element vertical and sometimes guyed towers. With the introduction of FM radio and television, *hill top* locations were employed because of the *line-of-sight* characteristics common to this broadcasting mode. Cellular telephone and PCS services require another antenna configuration still. In fact the development of personal communication technology has changed just about everything previously accepted as being typical for an antenna site.

Monopoles dot the horizon along major highways. Buildings in every metropolitan area are crowned with increasingly familiar side mounted directional antenna structures. These systems provide the expanding infrastructure for the personal communication industry. When cellular service first came on the air, two systems, (the "A" and "B" system) were authorized for each definable population center. The "B" system was to be operated solely by a *wire-line* telephone company (or partnership), while the "A" system could only be operated by a company or association not having telephone industry interests whatsoever. The "A" system generally wound up being a combination or joint venture of several competing non-wire-line applicants.

FIG 1.1
BUILDING WITH A CELLULAR ANTENNA "CROWN"

This competitive *two-system* requirement by the Federal Communications Commission (FCC) was thought to be necessary because of the high investment capital needed and to insure that cellular service would not be so outrageously expensive so as to keep it out of the hands of the general public. The first systems came on line in the early 1980's. Shortly thereafter the rush was on to get not only the top 30, 60 and 90 cities on the air, but all the areas in between. This *build-out* process involved the need for reliable service along with the ability to serve an ever increasing number of users. In both cases these factors, expanding territories and increased demand within a given area, have driven the need for additional antenna sites up to the present. Now with the introduction of new services the process is starting all over again.

Coverage problems emerged very quickly. Reliable service in tunnels, under bridges or in below-ground parking facilities had to be provided. It is not uncommon for a typical cellular system to be operating in excess of one hundred antenna locations with as high as 30% of the them dedicated solely to solving specific coverage problems. The key to good coverage is also the ability for a mobile unit to pass from one radio coverage area to another without service interruption. This process is called a *hand off* (the computer transferring of call from one cell site to the next).

1.2 The Impact Of The "No-holes" Vehicle Antenna

Most initial cellular systems were designed to use an automobile installed mobile unit or car phone. These units were comparatively large, were marketed by many manufactures, and contained attractive features such as automatic dialing, *hands free*, message taking and horn alert to name a few. The advantage of the mobile unit was that the highest permissible power could be utilized enabling greater range if the antenna was attached outside of the vehicle clear of obstruction. The vehicle electrical system provided for unlimited use. Permanently mounted antennas on vehicles are a common practice for police and fire vehicles. But it soon became apparent that this was not to be the case with the emerging cellular subscriber. Cellular installers argued until they were blue in the face the advantages of the roof top mounted car antenna. It clearly does provide superior cellular service, range and clarity. But the arguments for the most part fell upon deaf ears. As a result, alternatives were introduced such as the *magnetic mount,* the *trunk lid mount* and a new concept in antenna systems, the *stick-on, no-hole* or *thru glass* antenna.

This glass mounted antenna was the brilliant idea of a reputable antenna company and became an overnight standard in the industry. But the process of coupling RF radio energy through a car window glass (especially if tinted) introduced power loss. In fact about 30 to 50% of the power of a typical mobile unit is lost in the process. It was argued that this problem would require cellular companies to add additional tower locations to make up for the shorter range encountered. In reality other factors required additional antenna sites including the technical problem introduced by the no holes antenna. The disadvantages encountered with the wide spread use of these antennas actually paved way for another unexpected growth phenomenon, i.e. the low-powered portable unit.

Other objections to the mobile mounted telephone unit were obvious. It could only be used while in or near the vehicle. Peak periods during morning and afternoon drive time reeked havoc upon system loading resulting in circuit busies and delays causing customer dissatisfaction. Off-peak periods of the day had little usage. So systems had to be constructed to accommodate the peak periods with lessor returns on the investment.

1.3 Portable Units Replace Mobile Units

As the technology and system coverages improved, manufacturers were busy developing the hand-held portable telephone. Even though cellular protocols were effectively open, manufacturers closely guarded their implementation techniques to miniaturize their products. Some of the developments have been astounding. What was sought was an "acceptable" size, weight and battery life. Some of the early versions of hand held units (referred to as "bricks" and closely approximating them) are downright funny to look at today.

Few of the early portable units (now referred to as "handsets") were used. But not for long. Surface mounted component technology, multilayer printed circuit boards and the development of specialized multi-function integrated circuits (IC's) coupled with creative plastic housings, and true innovations in battery technology quickly captured the preferences of mobile users to the point that today the portable is the unit of choice.

What is less obvious but very important to you as an antenna site owner is that the portable power level is almost five (5) times less then that of a functioning mobile unit. The significance of this reduced power level requires additional *cell splitting,* thus more antenna sites. But it is a win-win situation because with additional sites, the cellular companies can serve even greater numbers of subscribers using the same channels. The engine driving all this expansion is obviously the increasing subscriber base and the increased ease and convenience of obtaining revenue from usage that is no longer dependant upon peak periods related to when users were in their vehicles.

1.4 What Happened To The Old IMTS?

In the early days the forerunner of cellular service was called Improved Mobile Telephone System, IMTS, a great name because many generations of improvements were made to the system over the years and the name never changed. IMTS and cellular existed side by side for a time but the real death for IMTS occurred when manufacturers stopped supporting IMTS products. There is a lesson to be learned here. Manufacturers will usually be the first to abandon a market which is not expanding. Attention to this practice can be a valuable prediction of what will be happening in the future.

Site Savvy

> Manufacturers will usually be the first to abandon a technology which is not expanding.

However, the frequencies used in the old IMTS service were not abandoned but reemerged when the Federal Communications Commission (FCC) assigned them for use in one-way paging applications. Many companies who saw their mobile subscriber base dwindling, now had a migration strategy to begin to develop one-way personal paging applications. New networks evolved which were being supported by a different product. Many old mobile service providers also became players or at least agents for the new cellular services. Only those who refused to accept the changes and adapt to them were hurt by the transition.

1.5 New Paging Services

The FCC wasn't finished with paging applications and about the same time released large blocks of new paging channels for licensing by existing and new telecommunications entrepreneurs. New offerings in the 931MHz. band were met with little initial enthusiasm. Not much product was in place to support the new service and there was much speculation that the very high operating frequency would result in serious coverage problems. It wasn't long until transmitters, antenna systems and paging units began to fill the needs of serious investors willing to construct systems in major metropolitan areas. Today all kinds of products are available for both the operator and the paging user. Cellular type multiple site technology was adopted quickly by the paging industry when it was discovered that it was possible to simultaneously *turn on* several transmitters at the same time. *Simulcasting* pays close attention to the accuracy of the base station frequency of each channel and the phase and timing that audio and digital information has when it leaves the transmitters. If everything is adjusted properly, the rf signals which overlap can actually enhance coverage in an area. The build out of paging has also been accelerated drastically in the last several years with the (precarious?) use of satellite control, making the installation of a *fill-in* station as easy as finding a site, aiming a dish and plugging it in. A typical *fill-in* location usually requires an antenna site which may not particularly be in a great location. For example, you might own an apartment building at the bottom of a hill surrounded by even taller hills. The general area may represent a *knife edge shadow* problem for the paging company. Their signal, emanating from an adjacent hill, glances off the hill overlooking your location resulting in poor or distorted coverage below. Unless your location is right in the middle of a major metropolitan area, in which case it is highly likely the problem has already been solved by the paging company, you might be able to provide a useful and profitable solution to their coverage problem.

Many predicted that as cellular and PCS providers offered value added services such as voice mail, call transfer and numeric messaging etc. the paging industry would dry up and blow away. But paging subscribers continue to increase with no apparent reduction in new starts. Communication analysts and financial investors have been scratching their heads over this one. Perhaps it is the changing nature of paging service itself. First used for business applications, statistics now support a rapid shift to personal use as a whole new generation of teens and family members join the ranks of new subscribers.

Early voice pagers actually broadcast the original calling party's message. But paging companies quickly learned that a typical paging channel could only provide service to about three or four thousand voice subscribers before real time constraints choked the channel with traffic. Talk-time was gradually reduced to just a few seconds barely enough time to convey a telephone number before being disconnected. With increasing demand for units, a numeric technology was proposed which could accommodate roughly fifty to eighty thousand subscribers processed on a single channel. None of this was possible until the rotary dial telephone was replaced by the *touch-tone* phone. With the move to numeric beepers, alphanumeric paging was not far behind. To encode a text page one either has to dictate the alpha-message to a dispatch operator who encodes it into the paging terminal or use a dedicated computer with special software. Pretty cumbersome at the time but the situation changed rapidly.

The transition from voice to numeric to alphanumeric is almost generational. Early users of real time voice paging preferred that service.

New users never really had the opportunity to use voice and thus feel comfortable with numeric paging. Many early variations of numeric paging used two-digit number groups called *data codes* to represent brief text messages similar to *ten-codes* used by police departments. Now that the popularity of the personal computer, the on-line service, the Internet Service Provider (ISP) and of course *e-mail* has invaded the daily routines of most of us, it goes without saying that the linking of alphanumeric paging to e-mail was the next logical progression. *TSR Wireless* and others offer systems that instantly allow e-mail messages to be directed to a text pager as well as to the subscriber's regular on-line service. Also provided are news headlines, sports, weather and stock market data flashed instantly to the text pager on a routine basis. The alphanumeric pager is also capable of receiving numeric pages directly from touch dial phones as well.

I have a friend who has a very sophisticated cellular telephone unit with all the extras, a one hundred (100) number dialing directory, internal voice mail and a host of other value added services. Could it be that this product, even to a technically minded user, may be a bit formidable to use or remember how to use? Perhaps such products are too complicated and less desirable especially if you must add all those extra services to your monthly mobile telephone bill. Paging, on the other hand is very simple.

One paging consultant has predicted that the world paging market might exceed 200 million units early in the century. The impact of this expansion in popularity is more product, cheaper systems and a host of services offered both locally and globally. In other words, services are going to have to get better to survive. Providers who can not expand their systems are likely candidates for acquisition by providers who will. All this means to you is more antenna sites will be required. Product development is fueled by consumer demand and the selection of new pager products and features is staggering. Even the large manufacturers are contributing new offerings constantly. One significant product release is the *Motorola RSVP* which provides a paging unit cleverly constructed into the battery of a *Motorola MicroTAC* pocket telephone. The pager does the usual numeric job with audio and vibrator alerting. The battery into which it is constructed provides energy for both the beeper and the portable. Truly a unique product which attests to the popularity of separate application services in a *bundled* world.

Technology also fuels changes in the way paging operations are accomplished. *SkyTel Corp.* has been developing a new paging data protocol making two-way paging practical. Initial efforts were met with a few technological challenges but the system is getting better with each new upgrade. An interesting twist to high speed digital transmission paging is the reintroduction of voice paging. This time the voice is digitized, compacted and transmitted digitally to the pager. Much faster than the old analogue method of transmission. Future paging fads might very well be voice service once again.

Other large companies are also providing messaging and acknowledgment services with a high degree of sophistication such as *RAM Mobile Data USA L.P.* and *Ardis*, a leftover venture from IBM's original nationwide maintenance system operating in the 450MHz. band. Systems are proposed, find favor then sometimes are overshadowed by competing technologies such as the *pACT, Personal Air Communications Technology* system derived from an Internet protocol and based upon CDPD technology. Competing with these protocols for two-way paging are many others such as *Motorola's FLEX, InFLEX and ReFLEX* systems. These methods will cost billions to deploy nationwide.

The players in this arena are definitely the "big boys" (see Appendix A) who have the muscle and the marketing base to forge effective nationwide one-way and two-way paging systems quickly and efficiently. When a system of this sophistication is deployed it typically follows the following phases:

1. Major metropolitan areas are first implemented using existing facilities (roof-tops) as much as possible to keep costs down.
2. Next, deployment is made along the major Interstate highways from one population center to another.
3. Finally, *build-out* is every area in between where a reasonable return can be expected with attention to enhancement and solving coverage problems.

Accordingly, it is only necessary to build a portion of a system in order to obtain a market presence and a positive cash flow. But the goal of *total area coverage* and reliability is close behind. If the product and service offerings are on the right track, system development then moves on to rural America picking up site after site until the marketing (coverage) map is "all one color" (showing reliable coverage). The concept of two-way paging in actuality is really *acknowledgment one-way paging*, i.e. the calling party knows the general location of the unit and is assured that the unit has in fact received the message. The real objective of this enhanced service is reliable page delivery. Another advantage of a transponder in a paging unit is that the system can locate the single transmitter to convey the page. Otherwise one must resort to broadcasting the page over the entire nationwide network in the hopes of signaling the unit. If the network can't find you, it saves your message until you show up somewhere and are ready to receive it. The advantage of this system is 100% reliable message delivery. *AT&T Wireless* has projected that about 2000 or more base stations would be necessary to complete a workable system on a nationwide basis. What is not projected but follows from past experience is that between five to ten *receive only* sites would be required to enable the extremely low-powered paging device to relay its acknowledgment. That would imply that for a nationwide system to operate effectively, roughly 20,000 receive sites might be needed. How many companies are going into two-way paging? At last count there were about six so unless these companies work together, we could speculate that over 100,000 receive sites would be needed in the next few years. Granted these sites will not be high quality high profile locations, but understanding what will be needed can assist your evaluation of what you have to offer.

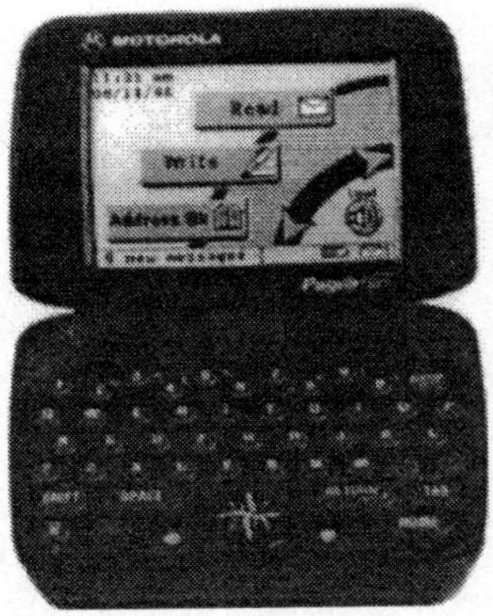

FIG 1.5
A MOTOROLA TWO-WAY PAGING UNIT

What about the demand for two-way paging? Recent studies have found that better than half of all paging users would like to know if their page has been delivered. There is no doubt that this statistic will drive the product and service offerings if the costs can remain competitive. What is equally significant is that many pager users, better than half of them, say that they always have to make a call back once a page has been received. These statistics were very different two decades ago when voice paging provided an extended message. It is obvious if all you can do is place a phone number in a pager's memory you're going to have to call that number to get your message. In the old days a voice message might sound like this "Joe, your three o'clock appointment has been postponed until tomorrow at 10 A.M.". It is not really necessary to call back on that kind of message. So what separates the functionality of a two-way pager from a cellular or PCS telephone? Price!

PageNet (Paging Network Inc.) one of the largest nationwide paging companies with a substantial customer base for numeric paging is developing its *VoiceNow* service to provide an alternative to the call back statistic, i.e. more information on the first page. In addition, there is a host of ancillary services that paging subscribers love to love. *ESPNet* provides sports scores and sport news features as regular broadcasting services to existing subscribers. Even more interesting, several FM radio broadcasters are paging on their sub-carrier channels making the lines between the broadcasting service and paging services murkier still.

1.6 Other Communications Services

At the same time that the FCC was creating the cellular industry, another service was catching hold in the 800MHz. band. It was initially referred to as *SMR,* Specialized Mobile Radio but now it is known as *ESMR,*, Enhanced Specialized Mobile Radio. This service uses multiple and trunked radio channels in a manor similar to cellular technology. Blocks of channels were granted for this service to be used primarily for fleet mobile and dispatched mobile applications. Struggling for standards and inter-connectivity, individual system operators hastened to develop dispatch services using telephone interconnection on a secondary basis. The service application was primarily for service or transport vehicles that require a dispatcher to tell them what to do and where to go. There are a lot of packages that need to be picked up or delivered and lots of plumbers, contractors and electricians that need day to day coordination with an office dispatcher. Although it would appear that these applications could be served by cellular telephone systems, SMR's usually invoice by the unit not by the minute. Smaller companies were likely to use cellular services but larger operations found SMR services far more economical. In recent years the basic dispatch aspect of SMR systems have been marked with some truly remarkable technological improvements, resulting in the blurring of the lines between SMR and Cellular operations. *Nextel Communications* leads the way in this endeavor with some truly creative services and are giving cellular and PCS services a run for their money. It is not uncommon, in any part of the country for former cellular mobile subscribers to be using SMR services to great economic advantage.

TV cable companies with broadband facilities are also in a unique position to provide wireless site services. But this potential has fallen short of expectations. Having an exclusive franchised territory (which is now being challenged by the Direct Satellite Service DSS providers) cable companies have a unique advantage of having a lot of cable hanging on poles, all connected to boxes which have power to them. With a minimum of special equipment it would be a short path to small antennas hanging from poles and attached to micro power bases stations, called *CMI's* or Cable Microcell Integrators. Networking their unique coverage footprint, cable companies are in an excellent position to exploit their embedded base and to provide Personal Communications Services (PCS). The bandwidth that most cable companies have going into buildings and residences is staggering when compared to just a telephone line. But all is not that easy however, as most cable systems are one-way and additional equipment would have to be provided to establish a two-way path to process PCS communication. And of course cable service is not provided to everybody. As exciting as this technological option appears to be, with its potential to make every telephone pole an antenna site, few cable companies have applied for PCS licenses. The primary advantage of cable operators in the PCS build out is that an area could be covered quickly and with much less capital compared to constructing tower sites. But the bottom line is the industry does not seem to want to wait around for cable companies to position themselves to provide the coverage which is needed. But don't feel sorry for the cable companies. They have a lot more service offerings in the wings such as providing basic telephone service, Internet access, high definition TV and home security monitoring along with their ever increasing subscription fees for basic television channels. We don't want them in the antenna site business anyway.

1.7 Services At 220 MHz.

In 1992, the FCC started accepting applications to be issued by lottery (now they use auctions) for several new service offerings and in particular one in the 220MHz. band. During this period "license mills" emerged using TV *infomercials* on non-prime time television touting get rich quick telecommunications opportunities. Many individuals who had little or no telecommunications experience invested in these applications in the hopes of "cashing in on the burgeoning telecommunications industry". The FCC indicated that there would be about 6000 licenses available for granting with over 4000 issued. It was envisioned by existing communications providers that this new service could have the potential of causing "destructive competition" a concept which the FCC rarely entertains much today like it used to.

Characteristic with any new service offering, are protests and entanglements. Issues were introduced which required clarification and resolution. The result of all these complaints and petitions for reconsideration sometimes causes the FCC to suspend the new services for a time. After the issues are resolved, the FCC releases construction permits to the winning applicants. Whenever a license is granted, the FCC is very serious about the applicant being able to construct its facilities and use it's assigned frequency allocation within a given period of time (usually one year) otherwise the license expires.

As an antenna site provider, knowing about these "hoops" will help you set up an antenna site agreement which initially allows a potential applicant to "lock-in" your site with an availability option. However, you are also entitled to an option payment for this privilege. If the application bogs down at the FCC or is challenged for any number of reasons (which is not your problem but effecting you nevertheless), it may be years before the option is converted to a monthly check. We will delve into this matter in greater detail in Chapter Seven. But for now, keeping abreast of new service offerings by the FCC, which seems to be happening more and more lately, can be a valuable resource in developing your site. Getting back to the above example, had you issued a site option to a prospective licensee in 1992 you wouldn't have started to receive a check for site rental until April of 1996 which was the absolute "drop-dead" date established by the FCC to have these systems completely operational.

In addition, applicant's have some incentive to drag their feet if engaging in a new service offering when investment capital may be at a premium or equipment suppliers are still trying to come up with a working product. Obviously a lot can happen in two years to a site. You don't want to tie up a lucrative position in your building or on your tower without being compensated.

Because site identification is necessary in an initial FCC license application, it is very likely that the site selected might no longer be available four years later. As a result, many of the construction permits might require modification or even the requirement to substitute other sites. Bad for you if you can't accommodate your original customer. Good for you if you missed out on the original application as you now have a crack at the business. This *Guide* will assist you in developing a resource to serve both sides of the equation.

The FCC addressed this issue when it granted its final *Report and Order.* It allowed antenna site moves within the top 50 cities to be less traumatic to avoid a new application or modification of the original filing. If the licensee was in any of the top 50 cities, the site could be moved up to 5 miles or 8 km. Note: the FCC likes to use the Metric System. If the licensee is in an area not near any of the top 50 markets, the site can be moved approximately 15 ½ miles or 25 km without any re-application for authority.

Keeping current on these twists and turns in the regulatory process insures your capitalizing upon them quickly. For example, all the potential applicants in your area are on file in the FCC Public Notices. In other words, you can easily find out who has applied for these services within 15 ½ miles of your site. If you know things are on hold because you keep abreast of the trade journals or the FCC Web site, you can contact the applicants with your offering and be ready to serve them when they receive their grants. Not all new services are met with such uncertainty at the onset. But it happens. When it does, confidences suffer. Applicants wonder if the application costs or option fees will use up important venture capital. Manufacturers are reluctant to commit to producing equipment which might not be sold for a while. It's a vicious circle. As with all new services, the bureaucracy grinds on and everyone eventually gets what they want at the end of the process. The cellular service, for example, took many years to resolve before the first licenses were issued.

Operations in the 220 MHz. band offer some interesting advantages which have yet to be completely explored. Some speculate that better in-building penetration and less susceptibility to absorption or other phenomena associated with cellular frequencies may be significant to the selection of this band especially for paging. One of the not so common uses suggested for the new 220 MHz. service includes an in-building robot that delivers mail and office supplies from room to room by computer remote control. Another proposed application includes point-of-sale devices which check credit cards, ATM and check depot services. Other suggestions include point of sale inventory and ordering systems, accounting control, state lottery sales points, sensing devices for toll gates and security applications. These uses are a far cry from the typical dispatch services initially envisioned. The new band is quite capable of supporting all types of services. The important thing is that the resource is a developing one and will no doubt result in many local antenna sites which you may be in a position to provide.

Keeping an eye on this industry should not be that difficult either as management consulting firms have already been providing construction and service advice to licensees in the form of weekly newsletters. In addition, some licenses may be up for grabs in less-populated areas. Industry notables have indicated that the market will indeed explode at least in the top 100 markets as other frequencies are becoming jammed packed with users. In addition the service was created at the cutting edge of data technology and will need little retro-fitting unlike the cellular service which is now in the process of reinventing itself in a digital domain. The important point about this new band is that you know it is out there and developing.

US MobilComm Inc. is a large 220 MHz. operator who is actively rolling out service offerings in the Washington, D.C. and Baltimore, Md. area. The company also operates systems in New York City, Boston, Philadelphia, Houston, Chicago, Milwaukee, Minneapolis, Miami and California. Since this system competes directly with SMR's operating in the 800 MHz. band, it is likely there will be *churn* (customers changing services) from one system to the other if prices for the 220 MHz. system can remain competitive. The company uses exclusive agents and some dealers to sell its present offerings. But remember as these services broaden and improve many additional antenna sites will be needed for expansion.

1.8 Band Selection And Antenna Location

The 220Mhz. band is an unusual band indeed. It has long been known that the major frequency bands used in government and commercial two-way radio exhibit unique coverage characteristics. Operating compatibility between bands is an important consideration and one for which the licensee is equally concerned. Should you have two tenants at your site, one transmitting a high power paging base station on 158.70 MHz. and the other proposing to use a two-way station with the mobile receive channel on 158.67 MHz., there is going to be an *in-band* problem. The paging channel will completely desensitize the mobile receiver channel when it is on the air because the two frequencies are too close for the receiver to block out the sideband energy from the transmitting station. This is a worse case scenario. However many incompatibilities besides adjacent channel problems can be skillfully managed by the tenants themselves. Let us look at some of the more general band characteristics.

1. Low Band, i.e. 35 to 45 MHz. offers greater over land coverage but has very poor in-building penetration. In addition it is effected by periodic sunspot activity occurring every 11 years or so resulting in *skip* communication. Skip is evidenced by transmissions bouncing down from the ionosphere, winding up about 2300 miles away and sounding just as good as a local transmitter. The band is not being supported by product and many operations are migrating to higher frequencies.

2. High-Band VHF, the 150 MHz. band is a pretty good

compromise as it is not effected as much by skip conditions and provides fairly good over land and in-building penetration. The problem with this band is that there are very few frequencies unassigned. Some proposals for splitting VHF frequencies to narrower band widths may add additional channels but unfortunately all the present equipment would require replacement.

3. High-Band VHF, the 220 MHZ. band is about the best compromise between overland and in-building penetration. It is unlikely to be effected by skip conditions. The band has not been used much for personal communications of any type and thus remains to be developed with new data services currently being proposed.

4. High-Band UHF, the 450 MHz. band is an excellent band for in-building penetration. However the over land characteristics begin to erode showing greater tendencies for line-of-site or point to point communication. Most of these frequencies are also in use with very little room for growth.

5. Super High Frequency, SHF, 800-900 MHz. and above, provide essentially line of sight communications. But one clear advantage of these frequencies is that they are electrically very quiet resulting in much less natural noise interfering with the reception of the desired signal.

Admittedly this is a simplification of the operating characteristics of each of the bands you are likely to see using your antenna site. But it is safe to say that your antenna site can accommodate services in each of the bands without much interaction. It is only when the services are *in-band* that care must be taken to avoid conflicts.

1.9 And Now There Is PCS

Should there ever have been a gold rush of the 1990's it would have to be the introduction, acceptance and construction of the new Personal Communication Service (PCS). With the Federal Communications Commission (FCC) exercising its new auction strategy to allocate frequencies for use in population centers, billions of dollars are being spent by very large companies for the privilege of competing with and cashing in upon the apparent financial success of cellular service.

What is the difference between cellular and PCS? The frequency range is different with PCS winding up in the 1.850 to 1.990 GHz. band. PCS also places a very large block of spectrum, some 120 MHz. in the hands of its operators as opposed to only 50 MHz. for cellular. PCS providers will start with digital systems using signal compression and a host of other computer permissible tricks to increase their capacity from 5 to 8 times over cellular. And by working in micro cells (that means lots of sites) the degree of frequency reuse can be vastly expanded. Cellular providers have a few tricks to maintain their edge for a time none the least of which is that they are already out there and doing a fairly good job of providing service. Many cellular subscribers own their equipment and may not want to purchase new equipment unless its dirt cheap or PCS service rates are significantly less. Remember that most cellular companies now literally give away their portables and mobile units just to have you sign a contract. So convinced are they that you'll average $75 to $150 month in service, they will get you started by giving you the phone just to get you hooked. But in this industry you are called a *POP* and giving you a free phone has proved to be an effective marketing tool. PCS equipment, at least initially, is going to be a lot more expensive. Many PCS players are already cash starved after surviving the licensing auction and the requirement to build a fully developed system from the onset. It is not likely they will be giving away their handsets initially. But no doubt we will be seeing all sorts of deals that will allow PCS operators to compete with the cellular companies already entrenched in the market?

With cellular annual revenues increasing at a rate of 35% percent or more, a lot of mistakes can be made without much impact. Some analysts project that within the next decade there will be over 120 million cellular/PCS subscribers. They are also projecting a new smaller service market of short range (home) based portable telephones. These phones may only work up to a mile from your house or office. If this was connected to a PCS system, there could be a huge new market for subscribers in a very brief period of time. What might even be happening here is a wholesale replacement of the *POT* or the Plain Old Telephone.

Unfortunately PCS is also beset with initial protocol standard problems. Not all systems will use the same method and interoperability will suffer without compatibility. For example, one popular system proposed for use is CDMA or Code Division Multiple Access. Another is TDMA, or Time Division Multiple Access. Finally GSM, or Global System for Mobile Communications is vying for acceptability. In some cases the protocols are identical but the frequencies are incompatible. So it may take a few years for all the systems to converge. Cellular on the other hand has already established a fairly reliable nation-wide system. Publicly traded corporations have spent unprecedented amounts at auction for a *block* of frequencies tied to a *BTA*, a Basic Trading Area and similar to the MSA mentioned earlier. There are six *blocks* of frequencies (A-thru F and not to be confused with the "A" & "B" cellular systems) covering the top US cities down to highly rural areas. Each *block* was awarded to the highest bidder in that market using a bidding process in which applicants wishing to enter the new service or market proposed a dollar amount per person within the population center.

The A and B blocks, already assigned and licensed for the most part, netted the Government roughly $8 Billion dollars in bids. Block C brought in about $10 Billion. That is a lot of money for just a piece of paper. Now the companies (see Appendix A) must build the networks, antenna systems and marketing mechanisms all capable of competing with not only two existing cellular systems and a few ESMR's already in place, but with other PCS operators as well. Someone described this situation as being similar to placing seven gas stations at a single intersection. Since many of the remaining areas and lesser spectrum blocks have yet to be constructed, there remains room for a lot to go wrong. Companies unable to secure financing for their awards or swap their *blocks* for more desired ones, will no doubt delay many market entries. Eventually the winners will shake out the losers and the systems will be constructed.

Communication companies are constructing facilities at the fastest pace in the history of telecommunications. Antenna systems including towers, shelters and switching equipment are going up in a matter of weeks. And the building process is not always the same between different companies. For example. When a PCS company reports that it has completed **Full** coverage, it means that it is approximately equivalent to cellular voice service. **Core** coverage means that initial construction covers the center city area or other core areas. And **Initial** service means that they are on the air serving "some" customers. **Full Commercial** service includes authentication, published pricing, billing and customer service. **Limited Commercial Service** means some of the above. **Trial** means they are still tweaking the system. These criteria, contributed by the *CDPD Forum*, give you an excellent idea of where the PCS operating company is positioned when they contact you about your sites.

In some instances the site they need may be desired immediately, or as they will specify, *time is of the essence,* in which case your deal is likely to go very fast. In another scenario they may be working a few months to a year ahead in their strategic planning and want to lock up your site for use six months down the road when they plan their transition from **Core** to **Full** service or from **Limited Commercial** to **Full Commercial Service**. In this case they are likely to drag their feet wanting you to commit and reserve the antenna site location but not willing to pay for it until construction is undertaken. Statistics as to where each provider is on this report card is a common reporting function of the various trade publications (see Appendix D).

Site Savvy

> PCS providers usually use buildings or existing towers first in order to minimize start up costs.

1.10 Other Services Are On The Horizon

There are far too many new services waiting in the wings or in partial states of regulatory or engineering development to write about in this *Guide*. For even as it is being written, new services are evolving and existing ones are in the midst of change. But the bottom line for you is that every service regardless of its sophisticated technology will still require an ever increasing resource of antenna site locations and a suitable physical environment to house equipment. Changes are occurring in conventional radio, the 220 MHz. trunking system, microcells, in-building cellular sites, wireless PBX systems, PCS facilities, one-way and two-way paging applications, along with a host of new product offerings such as wrist watch pagers, wireless laptops and palm sized Personal Data Assistants (PDA) *all interconnected in some fashion or another* by wireless modems.

Common to all this technology is *low power*. The trade off for these subminiature and low power devices will be to employ additional antenna locations. Rather than focus upon the technical aspects of these telecommunication services it is far more important to understand the physical requirements. You will then be able to match your proposed facilities to present and future building requirements of the communication company you wish to serve. In addition, you may even be able to anticipate facilities which the communication company may not yet have scheduled for improvement. You can do this easily by discovering an area of poor coverage. Essentially you must be a partner in these new ventures.

The *Antenna Site Operating Guide* addresses many different antenna site applications covering a wide range of individual activities or vocations from which you can benefit. You may want to skip to Chapter Ten and review some of these activities in greater detail in order to focus upon a given area as you work through the subjects in this *Guide*. Remember, you do not have to be an engineer. But you must do some research, assemble some information and develop a marketing plan to be successful. What makes it easy is that the communication companies you serve will welcome your suggestions whether they use them or not.

CHAPTER TWO

2.1 Discovering Antenna Site Opportunities

You own some high ground. You control a tall building located in a rural or suburban area. You have a structure of some kind on land that you own such as a water tower, a wind mill, a silo, some telephone poles, or even (if you are really lucky) a 100' radio tower. Perhaps you would rather speculate on the future value of property you are considering. Have you thought about the antenna site income potential of the land or building you intend to acquire. Then again you might want to lease a building and would like to be sure that you have the right to sublease the roof-top for antenna systems. Maybe you are an experienced real estate agent and would just like a lucrative side line activity.

Regardless of your motivation, this *Guide* explores antenna siting opportunities and the basics inherent, not guaranteed of course, in every property. Keep in mind that opportunity applies to both sites which are assumed to be excellent as well as to sites which at first glance may not seem so great. The *Guide* seeks to provide you with practical information so that you can evaluate site potential, construct, market and operate an antenna site for profit.

In this Chapter we will explore a few areas of practical site possibilities. Often a would be site owner's failure to recognize income potential or even appreciate the impact of the future growth of personal communications can simply result in missed opportunities. Have you ever said to yourself, *"if only I had bought those few acres or invested in that venture"*. Well my friend, that time is right here and now. You have in this *Guide,* most of the justification and information you will need to take advantage of an opportunity from the *ground up* (pun intended).

The following paragraphs summarize several specific areas of site potential which the author has encountered in his travels. No excuses are made for being somewhat cynical or critical in a few cases. For certain it has been the author's experience that there exists many potential site owners or managers who delight in saying "no" for the assertion. There are others who will listen intently but can not seem to reach a decision either yes or no. Finally there are those who listen, evaluate and act decisively. Often these three scenarios can be experienced in the same discipline. Study these applications carefully as they form the basis for many of the negotiation techniques to follow.

2.2 Real Estate Agents And Appraisers

Real estate sales people list and sell properties every day. Should a given property have some site potential, the agent should factor this into the selling price. Perhaps the agent can indicate to the buyer that according to the *Guide* criteria, the property may have some siting income potential as well. Notwithstanding zoning, local codes or any necessary construction to prepare the property for antenna site use, just about any property or holding can be effectively evaluated as to its antenna site potential. The author, in many presentations to real estate and commercial agents and brokers is amazed at the enthusiasm and questions which arise when agents are informed of what to look for in a property. Seemingly worthless right-of-ways or even odd parcels wind up having some really great potential for providing site locations if the real estate agent is cognizant and dutiful.

2.3 Architects And Civil Engineers

It is hard to believe how many buildings constructed within the last ten years could have been great antenna site locations. Because the designer apparently failed to chance any future applications even the most basic requirements necessary to accommodate antenna facilities are often omitted. I once located a brand new four story office building not only located along a major highway, but overlooking from a small hill, twenty (20) other single story buildings in an industrial complex. The complex housed several *Fortune 500* companies having thousands of employees all in view of this one building. The only access to its flat roof was a "pop-up" trap door about 2 feet square via a wall mounted metal ladder accessed from a utility closet. All the "stink pipes" emerging through the roof were of tin or PVC construction. Had the architect provided an elevator penthouse or a roof level superstructure of steel beams from which the main building was constructed, in order to mount antenna arrays; or, had he run some power and telephone conduits to the upper levels, a couple of 30 Amp circuits at 220VAC, a ground wire, a few coax cables and a 50-pair telephone cable or two; or, had he provided one small air conditioned room (about 20' by 20') as an equipment area; and, of course had he provided an easier way to get out onto the roof, the yearly income potential for this accommodation could easily have been in the $40,000 range or more.

Most architects will counter that the client didn't specify this accommodation. But did the architect even suggest it? Probably not. Of the hundreds of buildings which I have surveyed, eventually leased or upon which I have installed antennas, few had any accommodation designed for antenna site applications. The building in this case could not be used. In its present state it would require too many structural modifications to make the project feasible for the communications company who opted to construct a *single user* monopole on someone else's property nearby.

The building illustrated in Figure 2.3, however, included in its basic design, provisions for multiple antenna locations. And it is obvious they got a lot of business. Architects can easily provide preliminary site studies for their clients before the structure design is finalized. The long term income from a proper accommodation for antenna sites can help finance the building itself.

FIG. 2.3
BUILDING DESIGNED FOR ANTENNA STRUCTURES

Site Savvy

> When it's easy for a communication company to install it's antenna and cabinet, plug it in and hook up a telephone line, you will get the business!

2.4 Electronic Technicians

Ok, so you are a "ham" (an amateur radio operator) or perhaps you have your General Radio Telephone Operator License or are a certified electronics technician. Maybe you've been working for a communications firm all these years and you know a bit about radio and telecommunications. So much the better. Now it's time to cash in on your knowledge with your part time business on the side. If you work for the telecommunications industry you probably signed some sort of non-compete employment agreement that prevents you from going into competition with your employer. Would that you could. But what if you bought some mountain land which turned out to be a really nice place for your summer home or "shack" and it had a few extra acres to spare for a radio tower. If you are an amateur radio buff, you are entitled to set up a station for your own enjoyment. And when you do add a few extra yards of concrete to the base of your tower before you erect it. And oh yes, use a heavy duty tower because you might need to add some additional loading at a later date.

The technical information in this *Guide* might be familiar which puts you that much ahead of the game. In general however, you not only have the advantage of easily erecting a viable site but you also have many of the contacts you need to get it up and running. And because you already understand the engineering basics you need only review the marketing and operational sections of this *Guide* while offering your services (for hire) to other folks trying to establish and market their own antenna locations as a consultant.

2.5 Agriculture Land *(Antenna Farms)*

Look around your land. You might be surprised how close the *back 48* is to a major Interstate. So what if you are 50 miles from town. Those same folks driving along the Interstate, past your property, are also 50 miles away from town and more than likely their reliable radio coverage for a number of services needs improvement. If you are planning to sell the farm some day subdivide those *key* parcels. These are the ones on hills or with some right of way to power and telephone lines etc. When a *site finder* (me) drives a *search ring* looking for a quick-fix to marginal radio coverage (your place) , the real nightmare is to encounter miles and miles of fenced in land. If your road is a controlled access road, I might have to drive 10 miles to the next exit and double back to your land and then look for a house or someone who won't think I'm the uni-bomber or from the IRS when I ask "who owns" this land? But if you spend a little time checking your deed and your location maps, you might be surprised to see that "rocky knoll" is about 1000' feet from the junction of two major highways and could be prime real estate for a "build-out" project. Put a package together like the *Guide* suggests and send it to the folks listed in Appendix A. You will sleep a lot better knowing that your land has been marked on a telecommunication service provider's resources map. You might never get the call if you don't. And oh yes, next time you install a water tower, a wind mill, or a silo keep a few things in mind to help the customer get on the air real quick. Put up that sign too... ***"Prime Antenna Site Location Please Call. . . ."!***

2.6. Government Employees And Elected Officials

If you total up all the land which is owned and operated by local, county, state and federal government agencies it is astounding. I once found myself in an isolated little village in Massachusetts, a picturesque little place where the town hall was the highest building in the area. On top of the building was a "leaning" CB antenna that had long since succumbed to a wind or ice storm. In checking the topographic map, the town hall was ideal for a *microcell* facility and a paging fill-in antenna as well. Phone lines, easy access, everything was there except someone with the authority to say "lets get going"! A check with the town manager sent me to the engineer's office. There a young lady working as an intern for her PE (Professional Engineer) certification, said that "her hands were tied". They simply could see no way that *they (?)* could enter into a contract with any commercial company to install facilities on the roof of their town hall. "What about the liability issues?" I was reminded. That's why God invented insurance companies. On the way up to her office I snagged a brochure in the lobby describing how anyone, individual or company, could rent their public park pavilion for a picnic and a softball game. The concept is identical including the required *waiver's of liability* should someone twist their ankle sliding into second base or accidently spill some hot chicken barbeque sauce on their hands while cooking. I was amazed at the excuses given for not wishing to entertain *any* proposals for the use of the building even when it was explained that the site agreement was going to be for a minimum of 15 years and could produce enough revenue to heat and air condition the entire building for the term of the contract.

So if you are in government or a responsible public official, position your facilities for the requirements needed by the telecommunications industry. Much can be gained for your constituency. Observe the many telephone company and power company poles on government land. Do you suppose they are paying rent for all of them? A good place to start is a resolution which states: *our municipality will cooperate with telecommunications companies intending to serve the public* (need and convenience is good) *for the benefit of our citizens."* If it is too much for the government bureaucracy to handle, there are a lot of good site management firms who will coordinate the whole process and make sure a check is in the municipality general fund on time every month. You'll be a hero and will no doubt be reelected not to mention providing a few communications devices for your key emergency people in the deal.

To assist state and local governments with their wireless siting policies outlined in the Telecommunications Act of 1996, the Federal Communication's Commission Wireless Telecommunications Bureau has released a *national wireless tower siting fact sheet.* The fact sheet clarifies the jurisdiction rights of the local, state and federal government in regulating wireless tower siting. And is available from the FCC at its web site at **http://www.fcc.gov**

Sometimes our friends in government are so busy regulating things they forget that they are public servants. Recent developments have the pendulum swinging in another direction. Having been solicited by telecommunication site acquisition firms, some municipalities seem to have gone haywire in their attempts to not only control but to *steer* business their way. The introduction of restrictive site ordinances (we will review some of these in a later chapter) and zoning requirements along with other policies (some of them really strange) have created a fire storm of objection which in some instances seemly border upon antitrust practices. It is not uncommon for a municipality to "cut a deal" with a local telecommunication company. The company builds the tower and installs its antennas. The municipality installs its public service communication systems on the tower as well. Quid pro quo, right? Not much wrong with this arrangement. But then comes a requirement which suggests that the entire equity of the tower be conveyed to the municipality over a period of time coupled with another requirement for the company to build the original tower with *additional capacity* other than that required by the original two parties. And oh yes, the municipality "zones out" other suitable areas within its jurisdiction so that competing firms must first utilize the municipal tower at rental rates set by the municipality.

Some industry folks feel that a local government entity employing this practice is abusing its regulatory police power by asserting its dominant position in a supposedly free market. These situations have added months to PCS build-outs and have cost millions to adjudicate. Even more revealing is that if a new communication company entrant protests too much in the process, not surprisingly, other applications for construction may wind up in the "*push down file*" (a government term connoting punishment). But the companies are not totally innocent victims of this process either, as existing providers already entrenched in the market have been known to intentionally stir up the pot in order to delay a new company from becoming operational too quickly. This is one area of real sensitivity on both sides of the fence so it is important to recognize what forces are acting upon current decision making. When things settle down a bit there may be some really interesting court challenges to this process which appears on the surface to be in the public interest but in reality may be an abuse of due process after all.

And while we are on the subject of government agencies, quasi-public utilities and authorities are sometimes so entrenched in their bureaucracies that they forget they are in a very advantageous position to cash in on antenna site opportunities. Did you ever wonder why water tanks are on high hills? That one is easy. Water runs down hill, yet many water utilities haven't figured out that when a cellular company or PCS firm comes to them with a proposal to utilize their facility for an antenna site that indeed it can be done. In fact there are antenna mounting hardware firms which specialize in the crowning and structure engineering specially designed for using water tower facilities. In addition, more than one communication service company can be accommodated. The smart utility administrator allows telecommunication firms to make an offer and proposal along with supplying all the engineering requirements to insure a safe installation. Once again the additional revenue is a welcome margin to rate regulated utilities trying to hold the line on costs of plant and to satisfy bond holders.

FIG. 2.6
WATER TOWERS ARE USED AS ANTENNA SITES

FIG.2.61
HIGH VOLTAGE LINE USED AS AN ANTENNA SITE

Perhaps not as neat is the use of high power transmission towers for antenna sites. But it can be done if power company officials understand that its just another engineering accommodation. One problem seems to be lack of appreciation for the utility company right of eminent domain, and the fact that perhaps their facilities can serve more than one function. This includes not only structures but buildings and plant facilities happening to be in strategic locations. In some cases power utilities fail to take the initiative in seeking antenna site business. Most power companies are indeed in a very advantageous position to do so and the smart utility executive will create a Site Marketing Division, transfer a portion of its investment and retirement fund into communication company ventures so that it can be a "party of interest" and go after the business. Look at it from the telecommunication firm's point of view. "You have towers every 300 yards along a US Interstate...take your pick!" How can a telecommunication site engineer pass that up? But one thing for sure, this practice really bothers some municipalities and a few have even tried try to "zone-out" the practice.

2.7 Churches, Steeples And The Clergy Connection

One thing in common for most churches, religious institutions and schools, i.e. monasteries or convents, is that they are usually located in urban or rural areas, and from my experience, on high hills or excellent radio coverage positions. Then too, a lot of churches have steeples or bell towers. There are several companies specializing in *stealth steeple* antenna systems which can serve the needs of the cellular or PCS system in the community.

In a small Massachusetts town near a major industrial park and an interstate highway, there is situated a small church. It has a two story building serving as an educational facility. The location turned out to be at the exact center of my *search ring*. Coincidental? With a progressive board of trustees, realizing that perhaps there might have been some Devine intervention providing a partial solution to some of their financial problems, an agreement was worked out to provide a much needed antenna site. There are ethical considerations involved here. They need to be evaluated in detail by church authorities and members. But it seemed logical that there is no reason why a church can not benefit from its property when it can serve both its members and the needs of the community. Since churches are not in business for profit, the concept of a site license is no different than renting a church owned property or a fellowship hall/kitchen for a wedding reception. In my experience many folks in the telecommunications industry feel very comfortable with directing their resources towards religious institutions.

FIG. 2.7
A STEALTH PCS SITE HOUSED IN A CHURCH STEEPLE.

The antenna system has got be constructed one way or another. Using churches and church property such as cemeteries or church parking lots, assuming the tower structure can be tastefully designed and positioned, seems to have had few problems thus far. It is a working idea whose time perhaps has come. Church officials should explore their properties and present their findings to the industry in a positive manor.

2.8 School Administrators And Board Members

I almost wrecked my truck getting a picture of the Ft Collins,

Colorado Senior High School. I don't know any of the details but what we have here is a very progressive school district realizing the importance of using their plant for siting opportunities. This arrangement can provide income and the possibility of working with the communication companies on a continuing basis such as educational intern programs for technically and marketing oriented students. The picture and the situation speaks for it self. Other common applications for school districts is the replacement of lighting platforms on the athletic field with monopole antenna towers. This method provides an ideal, out of the way installation method benefitting everyone. But school districts either seem to be able to work with communication providers for the benefit of the community (including holding the line on increasing school taxes), or go completely berserk like the other example shown.

FIG. 2.8
CELLULAR ANTENNA SITE ON A HIGH SCHOOL BUILDING.

Here a school board authorized the installation of a PCS site on the roof of a multilevel elementary building. The deal was signed , sealed and installed. School board meetings were held, zoning was approved and building permits were issued. Then after a time, an objection was raised as to the possible harmful effects of RFR (RF Radiation) upon the children. The arguments advanced were emotional, subjective and unsupported by any scientific evidence. They were also originally advanced by someone not associated with the district, perhaps looking for headlines. Their objective was achieved. The local newspaper picked up on the controversy and a fire storm ensued. The school board members, perhaps fearing loss in a future election, attempted to back out of their contract. The communication company was willing to withdraw it even though legally it was executed and fully enforceable.

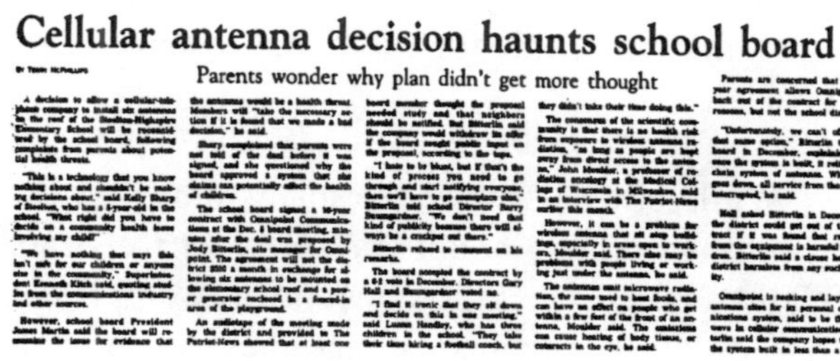

Cellular antenna decision haunts school board

Parents wonder why plan didn't get more thought

FIG 2.8 I TYPICAL NEWS PRINT WHEN CONTROVERSY ARISES

Who can testify that the reason "*why Johnny can't read*" is not caused by non-ionizing radiation emanating from this PCS site? Certainly no voice of logic prevailed in these hearings from my perspective. The building was old could possibly have had loads of lead pipes, lead based paint and asbestos all over the place (but "controlled" of course). Any negligible or measurable RF energy effects influencing *Johnny's* grey matter would also need to be compared with other deleterious influences such as televison, junk food, lack of exercise and parental interest in helping him with his daily homework. So you see, there must be some rationale and common base line for any decision by school officials to take advantage of this once in a life time opportunity for sustaining revenue other than unproven charges of possible brain damage. Here is another issue with which you will have to contend as you move forward. What I find curious however is that some school districts have absolutely no problem with the concept and others would take the matter to the Supreme Court if need be. If your school district opposes antenna siting, that is one less competitor for you.

2.9 Oil Delivery Firms, Electricians, Plumbers And Tradesmen

Medium scale firms of this type in business for at least 20 years are likely to have been pioneers in the use of two-way radio dispatched service. In many cases they have a free standing tower providing low band service which still stands unused today. Migration to other more efficient services in the 800 MHz. band have taken place years ago. Because the tower is a pain to remove, it is likely that it will remain in the air as the most convenient place to store it. If it has been cared for and is not rusted too badly, it may be usable for additional revenue in the paging industry as a receive site.

While site surveying the many suburbs of another Eastern city I frequently found one hundred (100') foot free standing (self-supporting, "SS") towers capable of holding one or two paging facility antennas. The beauty of these systems is that they are already there. If the local community now has ordinances addressing *new* tower construction, many of these structures have been "grand fathered" into staying right where they are. Some may still be in use for "control stations" but these towers continue to be marketable (subject to loading limits) and have income generating potential. In addition, many of these towers are located in land zoned as *commercial* so there might be room to house telecommunication equipment nearby.

I've been able to convince many tower owners in this class to let us rent space on their tower. But accommodating housing space for the equipment and providing the necessary twenty-four (24) hour access by the telecommunication company to repair or adjust its equipment has been another story. It seems like an easy problem to solve if one adds a shelter with an entrance outside any stockade area (the "yard") common to this type of installation. By design, small state police barracks with which I had some experience, provided a so called equipment room having it's own outside access door. Telephone and radio service personnel had no need to gain access to secured areas within the building after hours but merely accessed the areas assigned to them from the outside of the building. It would be a pity to loose several accounts because of the shelter accommodation situation. And oh yes, you don't need separate rooms for each telecommunications firm either. They usually use locked cabinets and are used to sharing floor space, outlets and other environmental facilities such as air conditioning and ventilation with each other. One secured room with easy access is all that is required.

Site Savvy

> It is not enough to provide space on your tower . . .you must also provide a location, with easy twenty-four (24) hour access to the telecommunication company for its equipment.

2.10. Contractors And Developers

High end housing and industrial complex developers take note. I discovered an area with marginal paging coverage and of concern to a client whose main facilities were a good distance from the nearest town. When I got into the area I found an absolutely magnificent residential area of very high priced homes. Several hundred of them in fact in the million dollar range, with six car garages, pools, and tennis courts. My utility truck was indeed relegated to the service entrance certainly not the main gate. To my amazement, the developer of the "Manor" had no doubt spent countless hours laying out the plots, the roads, lighting, services (cable power and telephone. . .all underground), even setting up a review board for architects, planners and contractors wanting to build in the development. But he forgot something important! As rolling and as pleasant as the surrounding terrain was, with spectacular views of nearby mountains, pagers and cellular telephones didn't work well at all in the complex. And there was absolutely "no place for an antenna" to be installed. Now, how long do you think something like that is going to last when a CEO of a large nearby corporation can't receive a page on his beeper, or a call on his cellular or PCS phone when he is playing tennis or in the pool? Not long!

The point is that with some planning, a utility location on high ground overlooking the complex could not only accommodate the required water pressure and natural gas distribution but also include housing facilities for antenna systems to be installed in either a well designed service area or if absolutely necessary in a disguised mode. Better still, each house in the complex could have been pre-wired with a special transmission line used in tunnels which would permit that structure to receive reliable coverage from some central point.

And the same goes for industrial complexes. The typical situation here is a large area of land upon which is constructed a series of single story offices or small plants. In some cases corporations may build their own facilities. When *light* manufacturing is permitted, there may even be a water tower or other high structure needed and nearby. When configured by the developer as a core construction project, consideration should be given at the onset for positioning of communication facilities. Communication companies won't want to extend their services to your complex until it is developed and rented. Then they want in! By this time it may be too late to add a communication facility. So, zone a tower and build a building to house the equipment <u>along with</u> the development. Position yourself for the lucrative and long term income to follow.

Developers also have lots of land, options and proposed projects in various stages of development. Keeping an eye on the site potential of these parcels early on may provide some outstanding opportunities for future income. Communication firms also love to deal with developers because they get roads and fences and footers poured quickly. So not only can you benefit from land rents but you may also pick up a good part of the site construction work as well.

2.11 Property Management Firms

With absentee ownership, large corporate real estate holdings, and trustee holdings in real estate, *property managers* are not far behind. Here are a group of building operational specialists, often certificated by national trade organizations and licensed by state real estate boards who "handle everything" from new leases to building maintenance. For a fee or commission, these firms completely manage any type of apartment or office building complex for their client/owners. In addition, the same company often handles a large number of properties in the same area.

Along with their conventional management and maintenance functions these firms often are the first to be contacted with regard to their buildings and what they have to offer the telecommunications industry. Typically if a communication company site finder knocks on the door with a siting proposal, the management firm is obligated to "take" the offer to the owner. If the owner is unconcerned, the process may take months, too long for a site decision to remain unresolved. Often, building owners, as individuals, family members or partnerships are content with having the building management firm take care of day to day operations. These professional services may not be the least expensive way to manage a building but they can certainly be the most reliable way to do it if a reputable property management firm can be found. When property managers understand basic antenna site operating guidelines, many opportunities to make additional revenue will be added for the client as well as the consultant.

When a site finder approaches a property manager, it is usually the maintenance man who is assigned the job of figuring out what is required. These folks rarely forward the advantages of the proposal to the property manager let alone the building owner. On the contrary, often it is dealt with as an annoyance. Absentee owners should add the antenna siting function to the list of managed services. Property managers should convince their clients of the worth of these endeavors. If the property manager is concerned about its lack of expertise in the area, specialized site property managers can be subcontracted to assist and train them. On the whole this can mean substantial additional recurring income to the property manager and the owner. However the process must be made swift and efficient or the deals will be lost.

2.12 Small Plants And Manufacturing Facilities

Areas previously approved for industrial, commercial and manufacturing carry a potential bonus should you be able to also add a tower for communications services. Unless towers have been "singled-out" of the process by municipalities, owners of these areas are in an ideal position to offer portions of their land for the construction of antenna facilities. Typically regional plans say that what ever is installed today will be able to remain in the future due to grand fathering rights. The construction of water towers, air conditioning stacks on building roof-tops or even a utility tower used for lighting can in addition to communication services be used by the plant and if preconceived, be augmented for antenna mounting opportunities. If the tower is constructed for security purposes, i.e. mounting a tv surveillance camera scanning your complex, the municipality can't give you much grief about it especially if they aren't doing a great job in law enforcement. Once the tower is constructed however, it can be used for any purpose you want. The point is to make some extra room for the future and reserve those potentially ideal locations on your complex for future siting applications.

2.13 Shopping Mall Owners

Here is one I love. All over the country sprawling collections of chain stores and rentals along with very large parking lots bring thousands of customers to a single site on a daily basis. More commercial communication company build-out decisions have involved poor coverage inside shopping malls than most any other commercial area. And it seems logical because the mall might also be located out side of town. They have gone through years of zoning hearings and development procedures to get going. Of the many malls surveyed very few have given wireless communication the slightest consideration. Not to mention that most cellular and paging companies are in fact store tenants as dealer companies or kiosk renters. How can you demonstrate the clarity of a cellular or PCS phone in a store when the coverage might be less than optimum? How hard is it to erect a big ugly antenna system at a shopping mall. Probably pretty hard. But just about every shopping center has a humongous shopping mall entrance sign.

With very little advance planning, the sign can easily house several telecommunication systems within its superstructure. In addition transmission cables can be "snaked" through the mall to allow specific coverages at all levels of the shopping area for telecommunication providers. Trying to get mall owners to consider antenna erection, after the fact, seems really difficult but it shouldn't be. There seems to be little logic why mall owners and operators can not meet telecommunication applications with reasonable speed and accommodation. When I see a communications monopole near but not on mall property, it perhaps serves to demonstrate a lack of forethought on the part of the mall owners, operators and investors.

2.14 Hospital Administrators & Medical Arts Building Owners

These facilities have long been providers of emergency medical communications and public service two-way radio systems. They were probably the first to use data applications with the wireless transmission of EKG and blood pressure information of patients in transport to the hospital. Many hospitals have their own internal paging systems and a host of medical personnel associated with the institution carry pagers, extension microcell phones and wireless telecommunications devices. Then too, many hospitals and medial arts buildings are located in expansive areas ideally suited for commercial siting. Roof-tops, air conditioning systems and ventilation facilities play a greater role in these buildings providing opportunities for siting income.

Many hospitals welcome telecommunication facilities not only for the revenue but for the convenience of the communications needs within the facility by both visitors, staff and patients alike. But there is a cautionary side to this application. Recent public scrutiny of RFR exposure has suggested that there may be some link between strong rf fields and interference with medical devices. But then again, power lines outside and internal wiring within the institution are said to cause the same effects. But what about electric blankets, computer terminals, coffee pots, microwave ovens or x-ray machines? Logic has not yet prevailed in these areas as well and we will deal with the subject in much greater detail in a later chapter.

How do communication companies deal with the RFR obstacle to siting in such public places as a hospital? Well it is not too difficult because there is no absence of medical opinion in support of these installations. The need for reliable communications during times of an emergency and the additional revenue are powerful incentives to provide siting accommodations. With proper design, shielding and careful antenna orientation all the RF fields are measurable on the simplest of instruments and any predictable effects can be minimized to far below any suggested nominal limits.

Final arguments for what is safe are long from being resolved but at present the power levels for these systems remains very low. Antenna systems used today are "directive arrays" with energy patterns aimed away from people. Nevertheless telecommunication facilities will more than likely be required to furnish RFR level data for all their installations in the future.

Folks responsible for hospital and medical arts building construction and operation are no strangers to the control of harmful radiation. They have been doing it for years with X-ray, laser, sound and magnetic devices not to mention construction, electrical and fire codes, chemicals, bio-hazzard and air quality matters within the plant. There may be times when the benefits of effective communication systems may even surpass the importance of compliance with guidelines for human exposure to radio frequency radiation within a medical facility. If the development of communications areas are a design criteria in the construction of new facilities most of these problems disappear. It is only the careful retro fitting of facilities in existing buildings which must be considered.

2.15 Fraternal Organizations, Retirement And Nursing Homes

Similar to hospitals and medical arts buildings are high-rise nursing homes, retirement communities and fraternal organizations such as *K of C, Masonic Temples, BPOE, American Legion,* and *VFW* or similar groups. Often these organizations are found to be in very favorable locations for siting applications but because of absentee ownership or committee management it sometimes is hard to get a quick decision. Telecommunication companies can't wait forever to get a decision on an institution's acceptance or rejection of a proposal. But one thing for sure, most of these organizations have a wide range of membership with differing backgrounds. It would be a simple matter for the reigning *Grand Leader* to set up a subcommittee consisting of an attorney, a communication guru, and a real estate person from the body of the membership to look into the prospect of offering the facilities for antenna sites. The revenues would be a welcome addition to capital investment and operations and the telecommunication firms will once again enjoy the benefits of your forethought, planning and evaluation. Not all of these facilities can serve siting applications but in my experience they are often way ahead of the game and worthy of consideration. In a community in Rhode Island, the highest location in a nearby suburb happened to be a Temple tower of the local Masonic Shrine. The organization is dedicated to serving the needs of disabled children. It did not take long to establish an acceptable arrangement for this antenna site when the proper committee was identified and approached. And like churches, once again, most telecommunication companies are eager to support local fraternal and senior citizen facilities in this manner.

2.16 Colleges & Universities

Established colleges and universities have been using two-way radio for security and maintenance of their facilities for a long time. Is it any wonder that many buildings of higher learning are in "lofty" areas often over looking small towns, villages or in downtown city complexes. These facilities are ideal for siting applications for many of the reasons above but can also serve as a class room for career minded engineering students to boot. The boards of trustees or directors need to explore the impact of these many siting applications so that when contacted they can be decisive concerning specific proposals. After all, I know of very few major telecommunications firms who do not recruit college graduates in marketing and engineering for their respective industry. There is no reason why the activity can not be of reciprocal advantage to the university. There is community college in Essex County, Maryland which maintains a huge friendly commercial site. It offers telecommunication, public service and emergency base station accommodations to wide range of entities. It is in so much demand that their main building is almost fully loaded and they are in need of providing a secondary location nearby. It helps that this college overlooks the community below. But there are other institutions however who are not so progressive. My own alma mater rejected repeated proposals for a credible communication company installation on a mountain top donated to the university. The difference I suppose is in the understanding level of the administration and as yet an "unarticulated objection". My university trust department can be assured that they will receive no donation from this graduate until the policy is reconsidered. (Take that!)

2.17 Hotel, Motel And Prison Administrations

While on a roof top looking at the horizon, I observed a brick tower jutting up from the ground about half a mile away. This place turned out to be a county prison facility located on a nearby hill top almost as high as the building from which I was observing. Prison officials had no idea that their location, with some minor modifications could accommodate antenna siting. Likewise hotels and motels have been used for years as excellent antenna sites because they are usually located along major highways, near airports, are frequently multiple story structures, have a good deal of space, roof area and user activity.

2.18 Volunteer Fire Departments, Police & Highway Patrols.

Independent emergency organizations and government agencies operating as boards or independent commissions and public service agencies often have need for improved transmitting facilities. Then too they always seem to lack the funds for the upgrades. The advantage of a deal between these types of organizations and a communications company is obvious. The public service entity provides the land and the telecommunication firm supplies the shelter, tower and the expertise to install the facility for shared use. Everyone comes out a winner. Often fire companies have existing structures which also have utility. As a bonus, telecommunication companies enjoy the twenty-four (24) hour security which often accompanies such an arrangement.

Fig. 2.18 Fire Observation tower serves as an excellent antenna site.

One possible large scale proposal yet to be popularized stems from the fact that the FCC recently made a large block of radio spectra in the 800 MHz. band available to public service agencies. State telecommunication *czars* wishing to upgrade aging low-band and VHF/UHF conventional two-way systems with trunked, digital and data services, were quick to tout the advantages of bringing not only state but county and local agencies into some new super public service network. But construction costs for some of these plans can be very high. One estimate for an entire state system started out at a mere half a billion dollars assuming that a reliability of 99.5% could be assured. Frequency resources don't seem to be much of a problem nor does the technology. The real challenge arises in the coordination of all those independent government entities, (referred to by a long time government friend of mine as *thief-domes*) distributing the costs for the project fairly and providing the infrastructure. The first two issues notwithstanding, a mature communication industry may soon or already be in a position to offer a complete infrastructure to the public service sector saving government millions in new construction and siting costs. What is it that keeps these parties apart? The introduction of subscriber use taxes, government interference in siting, lack of government communication officials thinking globally or perhaps just lack of trust on the part of communication managers when considering any long term relationship with government. Government has ways of funding everything it wants or needs such as the institution of E911 surcharges on local telephone service now being extended to wireless users. The bottom line is that the taxpayer pays the bill one way or another and its just a matter of how much the public is going to accept before the financial cart tips over. It would be a wonderful day if these entities could begin to work together at a level that *really counts* for the common good.

2.19 Radio and TV Stations

A peculiar problem has developed with the use of radio and some television facilities. This is a problem called a "hot" tower. Some broadcasting towers, even though they are ideally located on high hills or in remote areas are, as in the case of AM radio stations, the actual radiating elements themselves. If any other communication equipment is mounted on a tower of this type extra care must be taken to isolate the new antenna equipment from the existing radiating elements. Telecommunication companies are reluctant to use these facilities unless absolutely necessary. They often instruct *site finders* to disregard these towers as possible siting candidates.

Another not so evident characteristic of commercial broadcasting is that often the actual studios and the antenna sites are separated. Studio locations are ideal for co-location of facilities especially if the broadcaster is using a microwave link (STL) to send a signal to the remote transmitter site. Many TV stations are located on hills overlooking their markets and have all sorts of mounting opportunities. Usually these stations have an Engineering Director or "Chief Engineer" who can provide specific site information. Often the broadcast media will also be the recipient of marketing and advertising programs from the telecommunication carriers when they begin advertising their services and products affording even greater opportunity for cooperation. Telecommunication companies may need to rethink the use of broadcast facilities for their co-location however. Many manufacturers and consulting engineers offer complete solutions to the problem of any signal contamination.

Television stations who use heavy duty tall antenna structures may not be as amenable to siting opportunities now that many of these same towers might soon require augmentation for high definition television applications. There are some important opportunities here which will be addressed in a later chapter of the *Guide.*

2.20 Telephone Companies And Microwave Carriers

There once was a time, less than 30 years ago, when the phone company didn't want any competition. . .no way! There was a landmark case which tested whether you or anyone could mess around with a telephone at all without facing the full wrath of the telephone company's *foreign attachment* rule. It all started when a secretary wanted to attach a crude looking "hook" over the telephone receiver so that it could be held in place by squeezing the phone receiver against the shoulder, freeing up hands for typing. In addition another company, *Carterfone* wanted to insert the phone receiver into a cradle which amplified the ear piece audio to a speaker in order to provide an early version of a conference phone. Direct interconnection to the telephone line, common place today, was strictly forbidden let alone anyone hooking up an additional telephone extension, owning a telephone instrument or attaching anything to the telephone itself including telephone number stickers! Well, we have come a long way from those days.

One thing for sure most everyone pretty much knew how to *use* a *POT* (a "plain-old-telephone") too. Today, things are not always that obvious. *"Hey.....how do I get an outside line on this thing"?* What was not entirely appreciated was that telephone companies were also providing mobile telephone service and *"Bellboy"* paging. The FCC, when it created these services in the late 50's and early 60's, was quick to separate the entities in each market to insure that both a phone company and a "non-phone" company could be an area provider. I actually remember a public utility commission hearing in which some very learned attorneys attempted to grandfather "*tone-only*" paging (beep-only beepers) so that they would be exempt from rate regulation by asserting that in 1903 they installed a remote ringer (a slave set of telephone bells) in a barn so that if a call came into the main house it would also ring in the barn thus "paging the farmer" to a call. These arguments didn't provide many regulatory concessions.

This history illustrates the fact that because of the adversarial relationship existing between them, telephone companies and independent non-wireline common carriers rarely cooperated with each other when it came to antenna locations. Competition was fierce. Law suits and counter suits charged that phone companies were using telephone subscriber revenue to construct paging facilities thus asserting unfair competition with a non-wireline carrier who was left to its own financing to build its plant. And oh yes even a token ownership in the other's operation was strictly forbidden by the FCC. They were almost as adamant about that as they were about foreign ownership of FCC licenses, i.e. citizens of other countries could not own FCC licenses. Today, companies operate a lot differently. The burden of insider trading and interlocking directorships is pretty easy to establish or even to observe. But one thing remains the same, the shortage of ideal locations to construct antennas. You could always find the telephone company building in town. It was the one with several microwave dishes on it along with paging and mobile telephone antennas as well. In some cases phone companies even had aggressive real estate divisions whose job it was to actively purchase or acquire through long term leases antenna sites all over the place.

Cellular companies, whether owned by telephone companies or independent carriers, in the early 80's, essentially used the same strategy. Even though today competition appears to be fierce, it may all go away in the future as sites become more difficult to obtain and consolidations become more prevalent. It is all in the way the telecommunication planners view the scarcity of antenna sites. But the days of eminent domain and public utility access of land for antenna sites is pretty much over. That leaves a good part of the supply for future sites coming from the private sector. To be uniformed of this trend could cost you opportunity for profit.

2.21 Sport Complexes, Arenas And Ball Parks.

Thousands attend football games in major cities, near colleges or universities and thousands more ice shows other sports and music concerts. Yet few stadium and arena owners or managers offer their facilities for antenna siting. Empty for most of the time, these huge buildings or complexes usually have very high above ground locations offering ideal siting opportunities. Not only is there the height above ground, but many complexes are located along major highways, near major cities, have many tall lighting poles (if outside) and an abundance of equipment housing opportunities. These facilities rarely become primary sites but certainly can be included in build-out planning.

But *who you gonna call*? Often these arenas are managed by a commission, part time management or an authority. Rarely owned by a single person, they seem enveloped by their own entity with legal and liability clauses and definitions and regulations to the point if it isn't addressed it doesn't happen. When a site finder does knock on the door (finding the right door is often a problem) it will be months before the proper individual can be found, made to consider the proposal and further still until any deal can be made.

Stadium managers should recognize their siting potential and do some short-range planning and marketing. With the help of this *Guide*, arena engineering and maintenance groups can assist management in developing offerings which could be very attractive to telecommunication firms. The activity provides direct profit to arena operators not to mention improved event communication as well.

Did you ever wonder how all those corporate identity billboards, signs, flags and banners draped all over a stadium come to pass? It wasn't because the corporate sponsors, (many of whom are telecommunication providers and supporters like banks) "begged" stadium management for the opportunity to post their logos. It was because the marketing guys designed the format, regulations and spaces and went out and sold them. Well the principle is the same with siting opportunities. Perhaps slightly more technical but nothing a senior building maintenance supervisor can't handle.

2.22 Museums And Historical Locations

Historical sites serve literally thousands of tourists and visitors all over the country each year, often on shoestring budgets. There are many buildings, structures and locations involved all requiring parking facilities, gift shops, snack bars, camping and picnicking facilities etc. Signs point the way. Guides are standing by but no one seems to want to make a decision when a *site finder* comes to call.

FIG. 2.22 GETTYSBURG, PA OBSERVATION TOWER

A while back, I worked on a paging fill-in antenna site at Gettysburg, Pennsylvania. It's a quaint little college town. It is also the final home (farm actually) of President Eisenhower. But it is best known perhaps for the its battlefield and the Gettysburg Address. Although operated by the National Park Service, who incidently had no difficulty using their observation and fire towers as sites for their own two-way radio system, you would have thought that our proposal was a desecration of a national monument. The request would have improved coverage by providing public utility paging and mobile telephone services to the hundreds of thousands of tourists and Civil War re-in-actors assembling there every year. A separate tower was frowned upon as unsightly. Telecommunication companies were put through hoop after hoop in their efforts to get necessary ordinance clearances to erect them. Private property adjacent to the battlefield was the only solution and because the property was further and further away from the search ring, the towers became higher and higher (and more obvious). The power company had no problem erecting high voltage lines through the area

The logic for this whole scenario was further acerbated with the construction of a huge observation tower from which tourists could literally see the entire battlefield from a perch high up and over looking the entire area. The *Tower* was hotly contested from its initial construction, perhaps not so much as an "eye sore" but as a convenient way for tourists to get the whole visual and recorded three day battle plan from one location, for one convenient price, which included bathrooms, a water fountain and a gift shop, without going into town. Needless to say, it also provided a great antenna site. It was no longer necessary to hike through the paths and woods, getting eaten alive by the pesky Pennsylvania mosquitoes, black flies and ticks still inhabiting the area. Did I mention the poison ivy? This is not to say there are not those who enjoy the outdoors and the physical rigors of an actual tour. But there are those who simply are not up to it or prefer not to walk around all day looking at monuments. There are alternatives. A drive (stay in your car) tour with either a rented tape player lecture or a live guide who sits with you in your car giving you their lecture for a fee.

The fact that this the tower may have created competition with the battlefield ground tours not-with-standing, it was the original owner's intention to provide a very pleasant alternative to the traditional tours. Once again if the planning commissions, government agencies and local utilities had looked to the future from just beyond the next tourist, a controlled and very substantial antenna site facility might have been coordinated years ago providing an acceptable accommodation and along with producing a substantial financial benefit for the community. There is no doubt that the present mentality of *every company for itself*, as the number of single user towers scattered through the area will attest, does little to preserve the decorum of the area as may have been the original regulatory intent. Panting the towers "brown" another provincial regulation did little in my opinion to provide less obtrusiveness. And the truly sad fact was that a substantial part of the cost for these facilities was not the actual antenna system at all but the legal fees and procedural difficulties encountered to clear the way for construction. These "hidden" costs benefit the legal community for sure but had "marginal" impact on the end result. Perhaps there is another lesson which can be learned by these *battles* at Gettysburg. Managing the future may be a better solution than ignoring it.

2.23 Cable Operators

In general, and in the past, cable operators have been granted privilege to serve a community through franchise agreements or licenses. The nature of this business arrangement is changing drastically as rate regulation and competition with other carriers and (DSS) satellites alter the nature of the cable business. In return for a "one cable service to a community" (and the municipality's right to levy a franchise tax on your use of it), local ordinances seemingly restrict home owners from using alternatives such as C-band TVRO dishes (the big ones) and outside TV antennas which are "too large". These ordinances are commonly justified because the installations might be *adjudged* to be an "eye-sore" and accordingly, "not in the public interest". In the municipality where I once lived, you make an application to erect a dish but you'll need to go through the zoning board, do a survey of your nearby neighbors to see if they object and stand before a public hearing with engineering diagrams of your installation and grounding methods. You might even have to field a question (or two) as to your "need" for the dish. Incidently there appears to be no ordinance against storing an old refrigerator on your front porch so long as the doors have been removed. I wonder if a manufacturer could make a satellite dish look like a refrigerator?

Because Federal Law exempts Direct Satellite Service, (DSS) dishes from oppressive local regulation, cable companies have mounted vigorous anti-DSS campaigns to hold on to existing subscribers. The DSS deals are incredible even compared to cable offerings. However home owners and particularly condo owners are finding that to place a small 18" dish, even on the side of their house, opens them up to all sorts of easements, ordinances and obstacles imposed by the municipality even in the light of federal exemption.

Some have charged that this is restraint of trade, others have advanced arguments that because local TV stations are not carried on the DSS system local advertisers are excluded from viewing. Thus local TV should be forced upon anyone desiring a DSS installation. The local cable companies continue to raise their prices and "pass-through local franchise taxes" which might be difficult to impose when your service comes from 24,000 miles in space.

Fortunately the Federal Communications Commission has acted to clarify its 1996 rule preempting certain local regulations for satellite earth stations antennas the intent of which is to provide freedom of choice for the viewing public in the delivery of television programming. The action curbed a growing trend by local jurisdictions to ostensibly preserve its franchise tax base and its relationship with local cable companies through the imposition of "overly restrictive and unreasonable zoning laws". Section 207 of the Telecommunications Act of 1996 directed that the FCC preempt non-federal restrictions that "impair reception by antennas used in certain direct to home video services including direct broadcast satellite services". The FCC left a gapping hole in the process though. It excluded construction, electrical codes and safety matters from its exemption leaving the door open for local authorities to re-regulate these issues instead. The FCC's regulations apply to dishes of one (1) Meter or less in all areas, and two (2) Meters or less (a six foot dish) in commercial areas. Under the rule, local jurisdictions would be limited to enforcing only justifiable health and safety regulations. The FCC would decide any disputes only after "exhaustion of non-federal remedies" was made. The final outcome of this matter remains to be seen but thus far the FCC has pretty much favored the consumer. The attraction of DSS is growing and within a few years will no doubt take a significant market share from the cable services. But don't feel too bad for the cable companies, they own significant interests in DSS facilities.

FIG. 2.23
TYPICAL CATV TOWER

Now what has all this to do with CATV providers and antenna sites? Chapter One touched upon some potential uses for existing cable facilities. The cable companies actually have several areas of siting potential which they have been slow to release to the communications industry. A simple cable system begins with a very good antenna location (the "head end") usually consisting of hefty tower and a large building located on a high hill. CATV towers are easily identified as they have a host of TV type yagi antennas, any number of small dishes or corner reflector antennas mounted from just above the tree level all the way to the top. Because these systems are essentially *receive* locations, CATV engineers are not very responsive to site sharing their sites with high power transmitters because there is a chance of interference with sensitive wide-band amplifiers. Not so obvious is the fact that services like NPCS and two-way paging require "receive" sites as well thus lending themselves very nicely to this kind of facility benefitting both the CATV operator and the communication company. There are advantages in site sharing too. First and foremost is the shared FAA responsibility and liability for tower lighting and marking. Also, many CATV sales and service personnel use pagers and portable units obtained from the company who rents services from them. Finally there exists the possibility that the CATV company could also provide microwave channel links to the telecommunication companies. Corporate consideration of what is "acceptable" and what is inconsistent with CATV site operations are easy enough to develop. Trends in CATV operations favor satellite, rather than off-air systems along with a move towards using fiber optics rendering many CATV towers obsolete and used only as a back-ups. Site offerings for these facilities can provide additional opportunities for income with mutually beneficial advantages to all the parties involved.

2.24 Billboard Advertisers

Outdoor advertising companies are in a unique position to provide some of the best commercial siting opportunities to telecommunication firm build-outs in the country. With attention to the specific location and construction of certain types of billboard systems, stealth cell sites, PCS fill-in and paging applications abound. These facilities in some cases are very substantially constructed have adequate power, telephone lines and property with access to accommodate several subscribers. And because there are so many billboard operators around the country many smaller firms may not know what to do to get their locations in the data base and under consideration. As with other industries, some billboard companies are growing rapidly by acquisition of smaller companies. Also some companies have aggressive antenna site marketing policies. Deals are being made constantly and the industry continues to be a very lucrative source for road side communication build-outs. Installation of new billboards are sometimes met with a great deal of resistance similar in nature to tower construction. Quite frankly anyone traveling the Interstates for any length of time welcomes the food, gas and sleep information that enable traveling decisions to be made. In some cases the monotony of Interstate travel is also abated by outdoor advertising. However, when truly scenic areas are encountered, good taste usually prevails and the behemoth

FIG. 2.24
BILLBOARD ANTENNA SITE

signs are no where to be seen.

Nevertheless, if you operate an outdoor advertising company even if you just a have a few locations, you could be enjoying substantial revenue with the addition of antenna site functions to your facilities. When zoning outside advertising, the multiple use objective may or may not serve as additional justification for your venture. But if you can convince local planners that one less monopole would be needed because you have provided an alternative you have made a good argument. Then again it is very likely that one of your more sustaining advertising subscribers could very likely be the communication company using your billboard as an antenna site. What a sweetheart deal that would make. However what is required is a positive corporate acceptance of antenna site utility using outdoor advertising. Even when your board reads: "Rent this sign. . . call 1-800-555-1212" You'll be raking in the bucks and providing a very desirable service.

2.25 Home Owners

The largest single source of antenna sites are home and private property owners. Condominiums and other planned or restricted communities are usually excluded from consideration unless provided for *at the onset of the development.* Deed and zoning restrictions notwithstanding, there are many potential adjustments possible which can provide antenna sites under a wide set circumstances. Usually, communication companies are reluctant to waste time on these ventures but you certainly can give it a try. If approvals are obtained, the communication company will be happy to consider your proposal.

2.26 Other Applications Are Also Possible

We could continue along exploring many different applications but for the most part it is reasonable to assume that there are limitless possibilities for cashing in on the construction of antenna sites. There are literally millions of ideal antenna sites awaiting development. The more you can provide in your proposal the better your chances of being selected.

Start with the basics:

1. Is you property located in an advantageous position relative to something that you think might need improved telecommunication coverage.

2. Does your property overlook a factory or industrial complex or a highway where millions of vehicles might pass by every year.

3. Will your municipality permit you to erect a tower or telephone pole on your property and to "rent it out" as an in-home business. Remember, even residential areas allow in-home businesses such as authors, tax preparers, computer programmers, crafts or even an apartment or rooms for rent. Usually the operation of the business is directly related to the number of customers coming and going at various hours of the day or night. The less disruption along these lines the better. If your location can receive a waiver or can be operated as a commercial property for a small home business, you can expect your telecommunication client to make several installation visits and then you might not see them for six months to a year if everything continues to function normally.

4. Is it possible for you to provide a small, secure location with separate access for the communication company equipment? Most companies are not real thrilled about having their electronic equipment located in your basement or dining room. They require unlimited and unencumbered access to their equipment if they can get it. But sometimes that is not always possible as would be the case in a penthouse where building access is required. But in the case of a residential area, the more it looks like a TV cable, utility or telephone terminal box the better.

CHAPTER THREE

3.1 Site Identification

Every antenna site has to be located somewhere. Three site parameters will be needed to market your site: Height, Longitude and Latitude. Let us look at height first.

There are several points of reference we must determine:

1. There is the *height* above ground level, AGL; then,
2. There is the *height* above mean sea level, AMSL; then
3. There is the *height* above average terrain, HAAT.

If there is no structure on the property we must use ground level above mean sea level as our point of reference. If there is a structure on the property we determine the structure height from ground level to the highest point at which an antenna can be mounted and then **add** it to the ground level height at mean sea level.

To find ground above mean sea level you will need a U.S. Coast & Geological Survey, USC&GS, Topographic Map for the *quadrangle* on which your property is shown. You have several options here. The best one and least expensive is to obtain a "7.5 Minute Series Topographic Map" of your specific location. It will have a name in the upper right of the map such as "BENTFORK QUADRANGLE". In the lower right corner of the map it will say "Bentfork, VA." In general the community of "Bentfork, VA" will be shown somewhere on the map whose scale shows roughly 6 by 8 miles of land.

Complete information on how to obtain a map showing your property is available from the U.S. Coast & Geological Survey, USC&GS, Department of Interior:

U.S. Geological Survey		U.S. Geological Survey
1200 South Eads Street	also	Box 80225
Reston, VA 22092		Denver, CO 80225

It is not difficult to obtain topographic maps as they are also available by mail from **Carolina Global Maps, 1-800-321-6277.** Good quality stationery stores serving the engineering community sometime stock these maps locally. A call to your municipal civil engineer, a local architect or even a surveying company should get you the necessary information as to a good source for these maps. All maps are dated from the date of last revision. It is not uncommon for a map, if in a rural area, to be several years old since the last revision. When there are significant road changes, new developments or even improved satellite photo data revisions replace older maps. It is important to be sure you obtain the latest one. Also shown on each map is a *map key*. It can be identified as a small outline of your state with an even smaller black rectangle or *dot* printed somewhere within the border outline. The *key* shows the relative location of the quadrangle in relation to the area of your state. At each side, top and bottom and in the four corners of the map are the names of the adjacent or connecting maps. If you can find the map on which your property is shown you will also have the names of the surrounding maps as well.

Topographic maps are printed in several scales and sizes. We are interested in two. You will need one of each of them.

1. The 7.5 Minute Series map has a 20' foot contour interval and a scale of 1:24,000 (we will explain this later). This map is necessary to accurately determine both the ground elevation above mean sea level, Longitude and Latitude.

If your property is located near the center of your respective map you will need only one. If however your site is near an edge of the map you will need the adjacent quadrangle. The worst case is up to four maps if your property is located near any of the corners. These maps cost about $10 each depending upon where you buy them and how much postage and handling is involved.

2. A larger scale map, shows a lot more area. It has a contour interval of 100' and is referred to as the "1:250,000" map. This map can be used to calculate height, Longitude and Latitude but with less accuracy. We use this map to evaluate the *quality* of your site over a typical area out to about 20 miles from your location in all directions. We refer to this presentation as the *foot print*. It enables us to see any intervening hilly terrain or obstacles that may block line of site coverage from your site to other areas of interest. Should you live in the high plains and flat lands of Kansas this map will serve little purpose. However many states have some pretty rugged terrain. Accordingly, this larger scale map will supply much more information as to what kind of radio coverage *foot print* can be expected from your site.

Both maps have a *symbols* guide to assist in interpreting the locations of roads, trails, power lines, train tracks, boundaries, buildings, bench marks and the like. Some maps have detailed instructions printed on the reverse side as well, but not all of them.

All topographic maps are oriented to **True North** *not magnetic North!* A correction factor, a small angled diagram in degrees, shows the difference and vector between magnetic North to True North on any given map. Should you wish to orient the map to line up with a physical feature which can be seen but is not on the map you must use this correction to obtain a proper heading. As an example. You may want to describe a water tower not shown on the map as "2.1 miles at a heading of N233°E" from your site. You can't just use the heading given on your magnetic compass because it is wrong by the correction amount. And you can't spot the object on the topographic map without using a True North vector either. You align the map on a flat surface, using a magnetic compass. Then take a visual vector reading with your magnetic compass. Align that reading with the correction diagram for True North. Now shift the map slightly to the True North line. Finally take the vector reading with a *plotter* to determine the True North vector for the object. Some map suppliers may be able to provide a map user guide along with some tips and symbol sheets to assist in reading and using these maps.

3.2 Determining Ground Elevation

It is very important for you to *spot* your **exact** location on the 7.5 Minute map or as close as possible. If there is any doubt, due to indeterminate features on the map, check with your zoning officer. He will be able to assist you. Bring a copy of your deed along as it contains information which can be cross referenced to the topographic map.

Once your single point is marked on the map, it will either lie *on* a brown contour line or *in between* two brown contour lines. Chances are good it will be *in between*. Every fifth (5th) brown contour line is slightly **bolder** than the four lines in between. Each bold brown line, if you trace them around, will be broken somewhere and have a number inserted in-line with the contour. But as always there are some exceptions. The first occurs when the complete loop made by the brown contour line is too small to hold the number. Here you must assume it is either the peak of a hill, or lowest point in a valley. The other exception occurs when the brown contour line runs off the map at an edge; e.g. side, top or bottom, in which case the number you are looking for will be on the adjacent map.

Another example. Should your site be right on a bold brown contour line which is identified as "700", then your antenna site *ground elevation* is exactly 700' feet above mean sea level or **700' AMSL**. If your site lies half way between two lines but in the open area (not directly on a contour line) but say *between* 700' and the next normal brown contour line, and in a direction which, after four additional normal brown contour lines there is another bold brown contour line which reads "800", then your site is really between 700' and 720' AMSL. Remember, each contour interval is 20' in elevation or depression. Now by simple interpolation (spacial estimating) you can easily determine with a high degree of accuracy (within a 20' foot error) that your site ground elevation above mean sea level is 710'. It would be described as: ***Ground Elevation is 710' AMSL.*** Should you have a structure on the site such as a 60' elevator penthouse on a building, your site height would be given as an overall antenna mounting height of 770' AMSL, or 60' above a ground elevation of 710' AMSL.

Topographic maps are really quite easy to read. If you need some help try to find someone with military, aviation or civil engineering background who can double check your data.

3.3 Determining Your Site Location Using Datums

Positions on maps are unique locations such that one location on the earth's surface will not be confused with another. The fact is however, a single point on the globe can be described in any number of ways. There is for example the Universal Transverse Mercator (UTM) system. It is generally found on land based maps and quad sheets that are produced by government map providers. On land, this system is a bit easer to use than is Latitude and Longitude. However the model upon which it is based is a bit abstract. Think back to grade school when the geography teacher pulled down the *Map of the World*. All the continents, oceans and land masses were shown on one big flat rectangular map in the familiar pattern we all learned as kids. By drawing a vertical line through Greenwich, England and moving 5 degrees to the East or West of this line, you wind up with 60 zones (not time zones) to cover the entire earth between 84 degrees North and 80 degrees to the South of the Equator. The reason it doesn't go all the way to 90 degrees is that the polar area can not be described by UTM. Each zone is 6 degrees wide as projected from the Earth's center. Thus a UTM map is described by three elements, its zone, its *Easting* and its *Northing*. *Eastings* and *Northings* measure how far into a zone a position is in meters. *Eastings* are an East/West measurement and correspond roughly to Longitude (but very roughly!). *Northings* are North/South measurements and roughly correspond to Latitude. The problem with this method is that it is based upon a totally distorted view of the world essentially because it is being printed on a flat rectangular surface. Remember how big Greenland always looked? Or the fact that Antarctica was always spread all the way along the bottom of the map? We won't be using this method nor can we use any derivative of this map such as an atlas or a state road map.

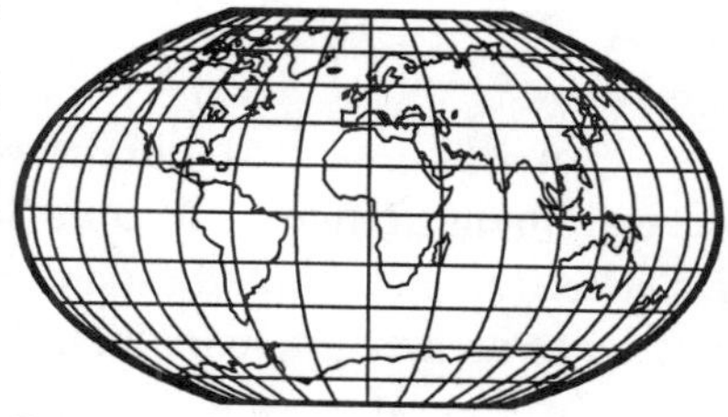

FIG. 3.2
UTM SYSTEM (SEE TEXT)

There are other map presentations as well. The OSGB is similar to UTB but describes only Great Britain. It must be used with the GBR36 Datum (survey benchmarks) which also describes Great Britain and it can't be used anywhere else in the world. If that weren't enough the GBR26 Datum *can* be used with Latitude and Longitude coordinates. And there are even more systems such as the Irish Grid, Swedish Grid or the Swiss Grid etc. There are also several different Datums too. The United States is now using NAD83, an update from NAD27. There is a World Geodetic System WGS84, an ALASK also known as the Alaska Datum, the GUAM Datum, the HUTZU (Hu-tzu-shan) Datum, the OEGYP (Old Egyptian) and the TANAN (Tananarive Observatory 1925) Datum to name a few.

In fact there are about seventy-four (74) Datums worldwide. That is a lot to consider when all you want to do is describe a single point on the the globe. The problem is that the Earth is not really a sphere at all but an oblate spheroid.

3.4 Longitude and Latitude Using NAD83 Datum

To obtain Longitude and Latitude we once again refer you to your 7.5 Minute Topographic map. We will be using the datum standard known as the North American Datum (NAD) which is based upon a geodetic standard for the earth's shape (ellipsoid). The first NAD was developed in 1927. Accordingly it was called the NAD27 coordinate system. It is the most common, most used and most understood. Because NAD27 was developed prior to satellite cartography we are in the process of moving to a similar but more accurate datum known as NAD83. This system was released in 1983 and is being used for the most part in all new map revisions. We will have more to say on these systems in a later chapter but for now it is important to determine what datum your map is using. Most revised maps after 1983 will be NAD83 but a few could still be expressed in NAD27.

An easy way to obtain very accurate Longitude and Latitude measurement from a 7.5 Minute Series topographic map is to use a special *DB Measuring Stick* made and distributed as a sales promotion by *Decibel Products,* an antenna manufacturing company. The special ruler works on many different maps and can be obtained from:

Decibel Products,
3184 Quebec St.
P.O. Box 569610
Dallas Texas 75356-9610
214 631-0310 Voice
214 631-4706 Fax.

Complete instructions come packed with this little plastic ruler explaining how to use the device on every type of map:

a..	USC&GS 7.5 Minute 20' Contour Interval	(required)
b.	USC&CG 100' Contour Interval	(required)
c.	USC&GS 15' Quadrangle	(not required)
c.	Sectional Aeronautical (1:5000,000)	(nice to have)

An Aeronautical map may be obtained for your area at most airports offering flight instruction. Even though the scale is larger than say (a.) above, it is ideal for locating the nearest airport and any other possible aviation hazzards near your site. It is also used to make a preliminary determination of whether your tower (or structure) will require FAA, Federal Aviation Administration, approval or lighting. In most cases however structures under two-hundred feet 200' do not require lighting or special painting if they are located far enough away from airport runways. We will deal with this subject later but try to obtain this map early on if you can.

It is also possible to use a regular ruler calibrated in tenths (10th) of inches or in centimeters (or any base ten measurement) to determine Longitude and Latitude but some interpolation will be necessary. The *DB Measuring Stick* will do the job very quickly and accurately because it directly reads in Degrees, Minutes and Seconds when you use the 7.5 Minute Series (or any) topographic map. Here is how it is used:

1. Determining Longitude using the *DB Measuring Stick*:

Initially you must locate the two parallel *longitude values* on the top and bottom of the map *between which* your site lies. In each upper and lower corner the full longitude is given in Degrees (D), Minutes (M) and Seconds (S). However, in between these corners it is only given as Minutes and Seconds with the Degrees assumed. There will also be "tick marks" every two (2) Minutes and thirty (30) Seconds apart. The total distance from one upper corner of the map to the other corner is 7.5 Minutes. That's how the map gets its name.

Once found, draw two light *vertical lines* in pencil connecting the "tick marks" along the top edge to the ones on the bottom of your map with your site location point some where in-between them. Remember that the distance between these lines will be some value between zero (0) Minutes and two (2) Minutes and thirty (30) Seconds. You must now interpolate (estimate by proportion) how many Minutes and Seconds lie on each side of your site location to the nearest vertical line. Both values obtained, when added together should equal two (2) Minutes and thirty (30) Seconds. This is a good check. Also remember that one (1) Degree equals sixty Minutes and one (1) Minute equals sixty (60) Seconds so don't forget to "carry" if you have to. Remember also you are working in geometry and not linear arithmetic.

If you want to use the value you have determined from the line drawn in pencil which is East of your site, *add* the value (the number of Minutes and Seconds) to the value for that line. Remember that the value for the Degrees is obtained from the complete number in the upper right corner of the map. You are only working to obtain the Minutes and Seconds. Conversely, if you choose to use the value to the West of your site, *subtract* it from the value of the line you have drawn and pick up the Degree number from the upper left corner of the map. In either case, the values should be the same. Be careful to carry anything over 59.9 Seconds to the Minutes value; and any Minute value over 59.9 Minutes to the next Degree value. If after you made your exact determination and if you obtain the same answer from both the East and then West interpolations you have done everything correctly and you now have your Longitude.

Get a feel for what you are doing. If you are confused, try to find a point already identified (benchmark) on your map and then determine its Longitude using the same method.

2. Determining Latitude using the *DB Measuring Stick*:

When determining the Latitude your pencil drawn lines will go from side to side, with your site lying within the parallel lines shifting North or South. The values will still be zero (0) Minutes to two (2) Minutes and thirty (30) Seconds but this time it will be added or subtracted from the Latitude values (and tick marks) along the side edges of the map. Using the same interpolation methods as used in determining Longitude, check your readings until you come out with the same value both above and below your site point. Now you will have your exact coordinates.

A few comments:

In North America it is typical to designate site coordinates with their relative position on the globe, i.e. above (North) or below (South) the Equator and to the East or West of Zero Meridian (Greenwich, England). Also, even though Longitude and Latitude are expressed in Degrees, Minutes and Seconds, less than a second, is sometimes expressed in *tenths*. See the example below.

Computers, have a tough time dealing with Degrees (360), Minutes (60) and Seconds(60) in programs which allow you to compute the exact linear distance between two sets of coordinates. These programs are quite simple and can be found in any geometry text book. You can also use a scientific calculator having keys for radians and trigonometric functions. However if the Longitude and Latitude can be expressed in only Degrees, with Minutes and Seconds being expressed as a decimal equivalent, computation becomes simpler still. Normally coordinates are expressed as:

Latitude North 40D 20M 21S
Longitude West 76D 52M 56S

But they are often expressed as Degrees, but in Minutes and Seconds expressed as a decimal (hundredths) of minutes. This can really be confusing. Take the Seconds and divide by 60 to get the decimal. Or take everything to the right of the decimal point and multiply it by 60. Example 20.35M would be .35 times 60 or 21 Seconds expressed as *20M 21S*. The same coordinates above could expressed as:

Latitude North 40D 20.35M
Longitude West 76D 52.93M

3.5 USING NAD83

The Federal Aviation Administration (FAA) has been ordered by Public Law 101-508 to use NAD83 (North America Datum) for geodetic coordinate information. NAD83 is based upon the Geodetic Reference System of a 1980 ellipsoid, which was "adopted" (perhaps not as widely accepted) as the official horizontal coordinate system of the United States. It is supposedly more accurate than ellipsoid models currently being used.

So effective October 15, 1992 the FAA converted from using North America Datum 1927 (NAD27) and any other horizontal datums, to using the NAD83 coordinates. All Latitudes and Longitude coordinates contained in Federal Aviation Administration (FAA) publications and correspondence after October 1992 will be based upon the NAD83 Datum.

Can you imagine the confusion? A tower having been identified on an FCC License for over forty years suddenly disappeared in a computer database because there was an error in converting from one coordinate datum system to the new one. I actually received aviation hazzard notices from the FAA for towers which have been in place and registered under NAD27 for decades. These towers, because their coordinates "changed", were treated as new towers when the FAA converted to NAD83. Some errors probably arose because the original site coordinate determinations may not have been accurate to begin with due to aging survey techniques. Combine these errors with the computer conversion program for the new datum, and "poof" towers disappear and reappear someplace else.

When dealing with the FAA, Latitude and Longitude in all submissions should be expressed as NAD83 coordinates even if you can only determine them using an older NAD27 topographic map. The conversion formula is not difficult. There is also a computer program available from the National Geodetic Survey, (probably in *Basica*) which converts coordinates derived from NAD27 (and other datums currently being used) to NAD83 coordinates. You can obtain a copy of the program "NADCOM" by contacting NGS at (301) 443-8631. Other questions concerning these systems can be called into (202) 267-9303 or (202) 267-9244 or at the time you make any filings, with any FAA Regional office.

If the coordinates submitted to the FAA are not in NAD83, the FAA may convert the coordinates to the equivalent NAD83 coordinates and all subsequent related correspondence shall reference the NAD83 coordinates thereafter. Recent FAA forms have a selection box for both NAD27 & NAD83 which may indicate that this differentiation continues to be an area for concern.

The justification to move to a new Datum is based upon the concept of *scientific obsolescence*. If the original system is relative to the new one, why change? For one thing, if satellite and laser technology has created a more accurate picture of the earth's shape, the change is fully justified. But some are of the opinion that the change was for change sake. As an example, there might be no continued justification for the continued use of the centimeter if the earth's shape has been redefined. And for certain a meter is no more accurate than a foot, a hertz is no more accurate than a cycle per second and a liter is no more accurate than a quart. They are just different units of measure based upon a different standard assumption. The National Oceanic and Atmospheric Administration, by a notice (decree?) in the Federal Register dated June 14, 1989, changed the official horizontal datum from NAD27 to NAD 83 for use in all federal government surveying and mapping programs as well. So the effect of this change also effects metes & bounds, lot & block and Range, Tier and Section methods of defining property boundaries. We may be seeing only the tip of the iceberg when it comes to the confusion these changes may cause. The new Datum NAD83, which is referenced to the Earth's center of mass, effectively redefines (changes) all Latitude and Longitude data used in the National Airspace System as well as including other modern world geodetic systems such as WGS 84. The horizontal differences between the old and new datums vary, ranging up to 350 feet in the continental United States with even greater differences in Alaska and Hawaii.

The impact associated with converting to NAD83 could be far-reaching. All files, data bases, charts, and publications displaying geographic coordinates must eventually be converted. Also the conversion was to have been completely effective on the same date (October 15, 1992) to "ensure that correct positional relationships" are maintained. The two Datums can not be intermixed. However, you may not be able to even obtain a NAD83 topographic map. An old map of any type will state: "National Geodetic Vertical Datum of 1929" somewhere in the lower portion of the map. It could also state: "polyonic projection 1927 North American datum", which is clearly an old NAD27 map. Using a revised NAD83 map initially will save a lot of time and confusion not to mention complicity with Federal Regulations.

3.6 Longitude and Latitude Using a Ruler

In the above determination we suggested that you use the *DB Measuring Stick* which enabled direct readings in Minutes and Seconds and thus was not subject to any interpolation error. However if you are unable to obtain one, you may use a standard ruler. I prefer a Metric ruler because it is calibrated in 10^{ths} . An American system ruler is usually calibrated in 8ths., 12ths. and sometimes even 16ths. and 32nds. of an inch. Occasionally you will be able to find a ruler calibrated in 10ths. of an inch. This rule will also work however all the others will be difficult but not impossible correlate. Since interpolation using *percentage of* is relatively easy, the Metric ruler, calibrated in 10ths or even 100ths. is the best choice. It will read-out in an exact number equal to the percentage you must adjust the Minutes and Seconds by, to obtain the actual coordinates.

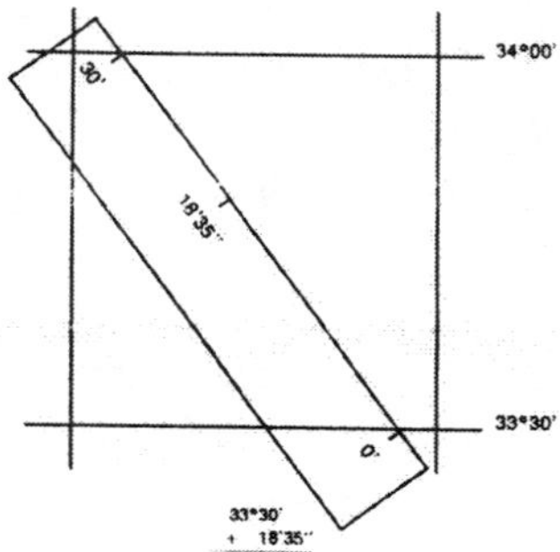

FIG. 3.6
POSITIONING CALIBRATED STRAIGHT EDGE TO DETERMINE COORDINATES

Remember, we are NOT using the rule to measure centimeters. We are using it as a simple proportional stick calibrated in tenths. The trick is to place the "0" value point of the ruler on one parallel line. The "10" value point on the other parallel line. Your ruler should now lie diagonally between the two parallel lines measuring 10 units between them for this to work. The ruler is further positioned up or down so that along its diagonal, the third point, your site, lies along its straight edge between the two parallel lines. That point on your ruler is the exact percentage you must add to or subtract from the values previously described along the top of the map. As a help, convert 2 Minutes and 30 Seconds (to all Seconds) or 150 Seconds, take your percent value times this number and convert it back to Minutes and Seconds by dividing by 60.

3.7 Software Methods

Fortunately there are some really neat software programs on the market today if you have access to a computer that can help you with your identification tasks. Here are a few of them:

Delorme Street Atlas USA for Windows© works really great for determining the Longitude and Latitude as well as making a very presentable map either in color or in black and white even using an older dot matrix printer. However, if your site is not in an urbanized area, it may be difficult for you to locate the exact position if no streets or other landmarks are showing. Also the program doesn't provide ground elevation.

Communications Engineering Technology, Inc. 1001 S. Ridgewood Ave. Edgewater, FL 32132-2332, 1-800-445-0297. These folks can provide every service you can imagine. Call them for an estimate of what you need. However you must still have your site coordinates for them to begin, otherwise they won't be able to find your site on their maps unless you actually send them your map with the location marked. They can provide cellular, paging and PCS coverage maps, contours (even in three dimensions), HAAT calculations and any other study you need. Although many of the services they provide are extensive and not actually necessary for your marketing presentation, they certainly can provide a detailed evaluation of your site so that to a radio engineer the data will be very convincing of what your site has to offer. The extra information if presented in this way can put you above the rest.

The extra information may be a good strategy if one of your sites has greater site potential for hosting a larger number of users such as an office building or high structure. Chances are also good that your first user will provide you copies of their engineering of your site whether they use it or not. You could always ask for a copy. It's worth a try.

On Target Mapping © 1051 Brinton Road, Pittsburgh, PA 15221 800-700-MAPS This is another company that will provide mapping and coverage programs and services for you with specific program "runs".

PhoneDisk© PowerFinder Digital Directory Assistance Inc. 6931 Arlington Road, Suite 405, Bethesda, MD 20814, 1-800-284-8353, provides telephone numbers and addresses for millions of listing around the country. They catalog by SIC codes for businesses as well as by name and address, telephone number and zip code. Once a specific address is located, (not an area) a detail option often brings up the Longitude and Latitude. I'm not sure how accurate it is but it gets you in the ball park quickly. It's a neat piece of software which is updated frequently.

3.8 Why Can't I use GPS?

Global Positioning System, GPS is a method of using satellites operating as a network to determine location on the surface of the earth. From a consumer's point of view, all that is needed is a small GPS receiver and some "open" sky. When you turn the receiver on it will eventually lock on to several satellites, perform some calculations and come up with the Longitude and Latitude in NAD83 and display it on the screen of the receiver. With built in computational functions, the receiver can also determine velocity, elevation, time in UT (Universal Time) and other directional and range finding functions. So why not just use a GPS receiver to determine the Longitude, Latitude and height requirements for your antenna site? It's wrong that's why.

For the most part, consumer GPS receivers get you in the ball park but they can't find first base. The reason is that the system, NAVSTAR, was developed by the Department of Defense for use in military applications. Because of the need for extreme precision for the military, consumer receivers may not have the same accuracy. We don't want some terrorist rigging a GPS receiver to a rocket now do we? The most accurate service is the precise positioning service (PPS) with a "published" accuracy of approximately seventy-two (72) feet horizontally and eighty-six (86) feet vertically. This may not seem like much of an error, but for actual positioning of an antenna site or property boundary it is not as good as what can be obtained using a topographic map. There is no doubt that some algorithms and multiple reading equipment can reduce this range of error significantly, but it is not what our government wants us to do. The operation of the NAVSTAR system proved to be a bit more accurate than it was supposed to be. So the Department of Defense actually introduces a random error which they term "Selective Availability" (SA) to maintain a military advantage. This means your reading might even vary from day to day if readings are taken at the exact same spot. The ephermeris information for each satellite is distorted in order to reduce its accuracy. It has been stated by DOD that "this error rarely exceeds 300 feet".

One method used to get a better "fix" is called DGPS for Differential GPS. It can increase the position accuracy to about 10 meters or 39 feet. DGPS requires that a local positioning signal be added to the "mix" of calculations made by the receiver. But these units operate at a much higher level of sophistication than we can justify for our needs.

Inexpensive receivers use a standard positioning service (SPS) which has an accuracy of about three hundred and twenty (320) feet horizontally and 960 feet vertically. Once again multiple readings, synthesis and plain old triangulation and interpolation can reduce this reading error significantly. But still not enough to specifically locate your antenna site for FCC licensing or FAA aeronautical hazzard purposes.

Therefore the GPS service, although becoming a very real part of the telecommunication scene, is just not good enough for precise site locating. Accordingly, the maps will have to do. Comparing GPS readings with measured (map derived) coordinates is a good exercise and certainly worth the time for experimentation. GPS does serve to provide *close enough* readings which can at least assist you in selecting the area you eventually want to precisely identify. It will also tell you quickly what map quadrangle you'll need to purchase. So it definitely has it's uses. But with the errors it introduces it can hardly be used without some other more accurate analytical tools to settle disputes, right of way or determine property lines, ground elevation and specific site locations.

What is interesting is how the GPS service is being marketed and applied commercially. You may want to take a few moments to explore this idea in case you come up with any good ideas yourself. A typical GPS receiver works a lot like a VHF monitor or hand held calculator. Its internal circuitry is very complex and can operate in a number of different ways. We mentioned earlier that the *Delorme Street Atlas USA* program was an ideal low cost map producing software. The company also offers a program called *Tripmate* which includes a GPS tracking antenna that plugs into the computer. Using the 24 positioning satellites in an orbiting *constellation*, the software finds your specific location on a map displaying it on the computer's screen. An indicator (a green arrow) points to your actual location on a map which slews beneath it as your vehicle moves along the road. GPS requires access to several satellites (a minimum of three or four) and *direct access* means antenna systems must be out in the open, not in buildings. The technology works fine for the truck or train industry because it can locate specific units along a normally preplanned route. Businesses who use this system profit from more efficient control and scheduling.

Summarizing, the accuracy as it is presently provided to consumers in the U.S. as controlled by the Defense Department is insufficient for our purposes.

3.9 Determining The Quality Of Your Site.

The final step in the evaluation of your site will take a little time. Remembering that most communication systems today take advantage of "line-of-sight" (LOS) propagation when transmitting high frequency radio waves, we need a qualitative way of judging just how good our site is. We need this information in order to attract the interest of the design engineer in considering our land, building or structure for inclusion in *their* future plans.

Visualize that we are standing at the very top of the tower, building or land and can look around 360 degrees in all directions. If we see the horizon in all directions, with no intervening hills (or valleys) the terrain will be pretty flat and generally will follow the normal curvature of the earth. This situation is common in the flat lands of the Midwest and Southwest etc. More than likely, however, we will see *terrain features* and man made obstructions and lots of them. As an example. If we look due North (not magnetic North) we might see gently rolling hills for several miles then perhaps the side or slope of a large hill cutting off any further view in that direction.

If there were no topographic features obstructing our view towards the natural horizon it would be a lot like what we would see when looking out into the ocean from the shore. We can only see "out" so far. If we observe a boat going towards the horizon, it will appear that its mast will "sink" slowly into the ocean. The mast does not sink for we are actually observing the effect of the *curvature of the earth* limiting our ability (aside from visibility due to weather conditions) to see *out* only by a fixed amount. The actual distance is approximately 35 miles but that too is dependant upon our height above ground and Latitude of the observation point. Continuing, look due South. We might see several hills some near and lower and some still further out. Maybe even hills that appear to be slightly higher than our location. To the West things might be very abrupt. We could be looking directly into the side of a mountain only a few miles away. The purpose of this exercise is to demonstrate the complexity of describing site characteristics because things can be very different in many directions. Our objective then, is to provide a *picture* of our site's quality, i.e. how the site will perform, for the prospective telecommunication company.

The Federal Communications Commission established a procedure in pre computer days which provides a fairly accurate graphic assessment of surrounding terrain. The procedure has evolved using the computer with a much greater degree of accuracy. However the older method can be easily developed manually from maps and will still serve our needs quite well. This presentation does not predict radio coverage. We will leave that to the telecommunication company engineers. It does however show the actual terrain, viewed from the side, in eight (8) (or more if desired) *radials* around the compass rose.

Communication companies are generally interested in two things:

a) Reliable coverage within 2 to 10 miles.
b) Extended coverage (or what coverage may result) after 10 miles.

This is a fairly arbitrary conclusion but we must start somewhere. The reason why they are interested in *close in* coverage is that your site must provide at least 97.5% (hopefully higher) radiation reliability within the area of design interest. If it is a circle around your antenna site with a radius of say 6 miles, then the coverage area would be about 12 miles in diameter. In an urban area this might be all that is desired. Should your site have a LOS (line of sight) to the horizon, say out to about 30 miles, the engineers worry less about the needed 12 mile circle. Rather they are concerned that *too much uncontrolled radiation* might escape the design pattern. This is one reason why so many antenna sites in use now are called *low profile sites* and the reason why telecommunication firms are not as interested in very high locations, i.e. *boomer sites*. But not to worry, high hills have their applications too and we will be discussing their value as well.

Higher elevation sites can still be used for low profile applications. There are methods which shape or direct the radiation patterns. One method uses highly directional antennas. Another tilts the radiator downward so that the main lobe (beams) of radio energy will not travel *out* to the horizon. By pointing the antenna radiation in a selected direction or using "half-circles" or " heart shaped" patterns, the engineer can customize your site to do the exact job required. Likewise many sites frequently have *built in natural barriers* or blocks to LOS radiation which can be utilized by engineers to their advantage. These characteristics are referred to as natural topographic barriers and can be used very creatively to do very specific jobs. The use of natural topographic barriers, allow what appears to be *a not so hot site* to fit perfectly (in the eyes of a design engineer) into an over all network. You just don't know. That is why you must not be discouraged if your site is blocked to the North by a very high hill. This situation might be just what the engineer is seeking. So now you can see why it is very helpful to attract attention to your antenna site by providing a cursory but qualitative presentation of the real world environment that an antenna mounted on your location would provide.

3.10 Determining Elevation Profiles

The map we will use to develop this evaluation is the 1:250,000 or 100' Contour Interval Topographic Map. This one map (if you are lucky and your site is not too close to an edge) will be the only one needed. After locating the precise position of your antenna site and marking it with a small *dot*, draw ten (10) concentric circles around your site using the *dot* as the locus or center point. A simple adjustable pencil compass or protractor available in the school supplies section of most variety stores will work just fine. Make the distance between each circle exactly one mile. Use the scale of *Statute Miles* found at the lower center of the topographic map. Make your circles using very thin (light) pencil lines. When you have completed drawing the ten (10) circles the outer circle should reflect a twenty (20) mile diameter around your site and ten (10) miles out or radius from the center.

Once you have the ten (10) one-mile concentric circles drawn, you will need a straight edge. A one foot ruler will be fine. You will also need a *compass rose*. A compass rose comes in several forms. A good quality unit made by HELIX (Model L09) is available at most drafting supply houses. It is a clear plastic circle about 6" in diameter calibrated in 360 degrees along the outer edge. Another really fine unit is made by Jeppesen Sanderson and is designated as the *PJ-1 Plotter*. Private pilots use them frequently and they may be obtained once again where you purchased your Aeronautical Section map. Even a chinchy plastic unit from a school supply store will work. These units are about four (4) or five (5) inches wide and look like *half circles* calibrated in 180 degrees.

Continuing, draw a vertical true North-South line intersecting the top of the tenth circle down through all the other circles to the site *dot* in the middle then on down to the bottom tenth circle. Be sure to align this line exactly North-South. To do this, find a tick- mark on the top and bottom of the map and make your North/South line parallel to a line drawn between these tick-marks. Your objective is to have your North-South line as *true* as possible.

Now, using the compass rose, line up the center point on your site and the "0" Degree indication on the North-South line at the top of the tenth circle. Mark off, clockwise, the 45 Degree (North East), 90 Degree (East), 135 Degree (South East) and the 180 Degree (South) line which is already drawn on the outer circle. Keep going around the outer circle. Mark the 225 Degree (South West), 270 Degree (West) and 315 Degree (North West) points.

Finally connect the 45 Degree and 225 Degree points with a light pencil line through your site in the center. Next connect the 90 Degree and the 270 Degree points through your site point. Finally connect the 135 Degree to the 315 Degree points. The resulting diagram should look like a *spider web*. What you have created are eight 8, radials, each ten (10) miles in length emanating from your site around the compass rose at forty-five degree intervals. So far so good?

Using a ruler calibrated in 16ths of an inch, place the origin (the 0 point) of the ruler on your site *dot* positioning the straight edge along the North or 0/360 Degree line which intersects all the circles. At each 16th. point on the ruler, read the ground elevation similar to how you did it in Section 3.2. Record the values on a previously prepared ledger work sheet. Now do the same for the other seven (7) radials. The first reading for each column on the table will be the ground elevation (not the structure height) of your antenna site and will be the same for all eight radials. When you are finished you should have eight columns of numbers, one for each of the eight (8) radials with approximately 160 readings per column. Yes it will take some time but do it right and you will have a very worthwhile exhibit to include with your marketing information.

Once all of the readings are obtained and entered into your ledger, total each column and divide by the number of entries (160) per column to obtain the *average ground elevation above sea level* for each of the eight radials. You may then add your structure height above ground to this number to give the Height Above Average Terrain, known as HAAT for each vector. Finally, average all eight HAAT values for the generalized HAAT for your site.

Don't be surprised if this number is small or even negative. Typically an HAAT of say 230' would mean the same thing as having a 230 foot tower on a flat piece of land 20 miles in diameter. The cost of building a 230' tower could be as high as $100,000. Even greater if it were self-supporting or designed with heavy loading in mind. As can be seen, the surrounding terrain (average) can contribute significantly to the engineering considerations. Here is another example. If your building is only fifty feet high (approximately a three story) but situated on a slight rise whereby your HAAT was 230', then the ground average from your site would be the same as your building sitting on top of a 180' tower. If your building is in the clear, i.e. no other taller buildings close by and blocking the ten mile path of the specific radial, the communication consultant might be very interested in talking to you about putting an antenna on your roof top. This information is good summary data. Now you must present the data graphically for even greater impact.

Look over your ledger of values. If any elevation entry value *is greater* than the calculated value of your overall HAAT figure, it may represent a natural topographic barrier blocking coverage of your site in that direction, at *that* distance , by *that* amount in feet. Likewise a value that is much less than your average amount could represent a *hole* in the coverage having the same effect. Typically this would be indicated by a fall-off in elevation along a radial such that the site height would *over-shoot* coverage into a valley. Much can be learned from an analysis of the data you have prepared.

In order to present this *terrain profile data* more clearly we merely graph each radial. The vertical axis of the graph is in feet above sea level and can start conveniently at any value lower than the lowest point you have read proceeding in a linear fashion to any value slightly higher than your highest entry. The horizontal axis is distance from 0 to 10 miles. The end result are eight line graphs showing the vertical profile elevation along each radial as if the earth were sliced into eight pieces of pie and viewed from the side.

FIG. 3.10 PROFILE GRAPH FOR ONE RADIAL USING *DELORME TOPO USA*

As simple a technique as this is, it is more than enough to capture interest in your site. Should you want to put more into your presentation you could order a computer generated terrain map, calculated at every degree, for as *far out* as you are willing to pay. Another possibility is to try ***DeLorme Topo USA***, an offering which permits cross sectional graphing on a given vector using topographic maps for the entire USA on a 4-CD-ROM set. It is an interesting application which may be suitable for this presentation. Some computer *runs* are expensive but the output can be quite impressive especially if printed in three dimensional color.

3.11 Three (3) Actual Antenna Site Models

Now that you have a basic understanding of how to identify what may be of interest to telecommunications planners and some skills in actually describing your own location, it is time to move on to some examples of actual antenna systems or "farms". Three (3) basic systems are presented. The first system is a simple fifty foot free standing tower providing for approximately four tenants.

The next system describes an intermediate site whose emphasis is upon housing equipment rather than tower construction. In fact the system is about as economical as one can get. It uses a single cinder-block style building and telephone poles for the antenna mounting arrangement.

The final system is a simple roof-top location.

These systems have some unique problems associated with them which will be discussed in detail. Structural integrity of the roof surface and nondestructive mounting arrangements must be provided. Access is always a problem along with special considerations providing for aesthetics, safety, security, electricity and telephone lines. None of these problems is insurmountable. However it is important, when laying out your antenna site plan to consider the potential growth of your site from the start. Allowing your first tenant to position equipment and antenna arrays to suit themselves usually results in the next tenant being compromised as to where he may locate his equipment and antenna system.

Chapter Ten offers additional systems and suggestions using the principles and products outlined in this *Guide*. Obviously the key to an ultimately successful site situation is **location**. If you have an area which is scarce and indivisible or if you have been granted zoning to operate an antenna site business, you are definitely in the *cat-bird seat*. It is not uncommon to have an antenna farm or single location capable of generating annual revenues in excess of half a Million Dollars. With a plant investment of less than $200,000, it is not only possible, but realistic to expect revenues of this magnitude but only if you are serious about the activity and stay in it for long haul.

3.12 A Basic Antenna Site

The assumption for this first suggested basic antenna site is that it is located on a high piece of ground. It does not have to be a mountain top, although that would be suitable in most cases. The hill, knoll or high plain must overlook a surrounding populated area and be within only a few miles of the heaviest population concentration. The area of coverage should also include a few well traveled roads, hopefully some industry, businesses or residential areas. The horizon should be free of obstructions and be at least eight to 15 miles distant and clearly visible. A location of this type is of general interest to a wide variety of communication companies. Typically a setup of this type should cost under a Thousand Dollars ($1000) and be able to support about four systems. If each system generates about $150.00 per month in site rent, it should be possible to recover the site development cost in about four months. What this means is that after four months of receiving $600.00 per month in site rents (subtracting electricity and incidental costs) you should be receiving a profit. Not bad for four (4) square feet of land. Yearly revenue after build-out and the construction cost is paid off should be about $6,600.

Fig. 3.12
Small four system antenna site

Other considerations are less obvious but equally important. Can the site location be along (but not too near) a major Interstate or thoroughfare. This is not only to provide highway coverage but so that it can be seen. If it is hidden away on a hill or among trees and accessible only by a dirt road it is less likely that it will be found easily.

The tower sections suggested for this system are light duty ten foot (10') sections. The overall height should permit the top of the tower to be at least thirty feet (30) feet above a hypothetical *tree-line* elevation. The *tree-line* consideration varies from place to place but for purposes of this example it is assumed that the average height of nearby trees will rarely exceed twenty feet (20') in height. In some cases, on high hills, the trees may rarely reach twenty feet (20') due to the rocky soil and lack of water. When vegetation is referred to as "scrub", it rarely needs to be a consideration in tower height. However in some parts of New York Pennsylvania and the Northeast the *tree line* average could be eighty feet (80') or more. *Tree line* refers to not only what is observed but what could happen in the next twenty (20) years. Many towers have been erected not taking this factor into consideration. Typically microwave shots become blocked, the tower itself becomes hidden and access may even be impeded. In some cases a taller tower can easily be justified before a zoning board by stating that the height required will provide for future tree growth. A little planning at this point and some observations will avoid any future problems.

The basic system, illustrated in Figure 3.12 is very near several outbuildings including a garage. It is also along a country road. The tower is visible for miles in all directions. It serves four (4) customers. Two services are known as *community repeaters* operating in the 450 MHz. band. There is a taxi service operating at 150 MHz. and a local paging company using the site for a remote "fill-in" location operating in the 932MHz. band. The system has little room for expansion but clearly provides our basic self contained system example.

Each tower section comes in ten (10') foot lengths. They bolt together at the overlapping seams using six stainless steel hex heads, lock washers and nuts, two (2) for each leg. The *smaller pipe points up* while the tower sections having the telescoping larger pipe, drops down over the smaller pipe. Construction of this system begins with digging a five (5) foot hole in the ground which is approximately three (3) feet square. The first ten foot section is centered and plumed into the hole. Concrete is then poured into the hole forming the tower base. The weight of each section begins to accumulate as more sections are joined pressing down on the connection points. Each section is hoisted up over the mounted section and "dropped" into place then bolted. The first twenty or thirty feet may be assembled on the ground, "walked up" and then rope guyed until the base drys. It is just as easy to securely install the base first then use a *gin-pole* to mount each additional section. In addition to the ten (10') foot sections, there are other sections and mounting accessories which will aid installation. Refer to the catalogs of the manufacturers for these options, mounting techniques and recommendations that will fit your specific applications.

A special top section, which curves into a pipe at the center, typically is used to hold a single smaller pipe in place upon which a television type receiving antenna and *rotor motor* may be mounted. We don't need this section in our case because you may wish to add a few more sections at a later date to increase the tower height. If the special top section described above is used, no additional sections can be mounted on the top. If additional sections are added it will also be necessary to consider using guy lines if the tower height increases much beyond the fifty foot (50') foot level. Close attention to the manufacturer's specifications will answer most of the questions with regard to this type of mast. The top pipes must be capped to prevent water from entering each leg freezing from the inside and cracking the pipe.

The important thing to remember is that this mast is of TV grade and is extremely limited as to its loading, i.e. the number of antennas which can be mounted on the mast itself. Major communication accounts will not be impressed at all by this installation. Nevertheless this facility has affordable utility for small local radio service applications and companies.

Because the system is a self-contained one, we must also provide power and telephone lines. Single phase 110/220 VAC at about 100 AMP service would be ideal. An eight (8) pair telephone cable is also required. The ideal installation for these hookups is underground in PVC conduit. Be sure if running services underground to bring conduits well up the side of the tower to prevent ground water from draining into the conduits or when snow levels might reach the entry point. The power may be metered for each client and should have breakers located at some point accessible to technicians servicing the equipment. Electrical service may be included in the site rental fee in which case separate metering would not be necessary. We will discuss the advisability of providing power in the next chapter. If it is necessary to charge separately for electricity, separate single phase power meters and bases will be required. These meters need not be solely provided by the power company. They can be installed and "read" by you and re-invoiced to your client.

Be careful with the power situation. Some services use very low current during receive modes but several thousand Watts during brief periods of transmission. If the average power use factor is small it may not be worth the effort of separate meters. However if you provide service to an FM broadcaster (broadcasting all the time) or a paging company with a *high traffic to power ratio*, you could be in for substantial power bills which should be paid by your client.

If you provide no building to house equipment, all the facilities used will have to be stored in *outside* cabinets. Usually the communication companies provide these cabinets. They are not as convenient as having a building. Outside cabinets are sometimes used as target practice by thrill seeking shooters. They can also be buried in snow . Finally they often pick up all kinds of "beasties" in the summer such as wasps, spiders and snakes. In the winter you can expect mice. Nevertheless outside cabinets are sturdy and in a small equipment configuration work quite nicely and are more than adequate. The example shown uses pole mounted nineteen inch cabinets side-mounted on the tower.

Two outdoor cabinets are mounted back-to-back or stacked using "U-bolts" to the tower legs. Be sure to mount them about four feet up from the ground. The power line and telephone *drop* may also terminate at the tower. If run over head make sure they are at least twenty feet above ground using proper hardware. Remember a truck snagging the power or telephone line could not only disrupt service but in this case pull down the entire tower. Try to avoid running service drops over roadways.

You should either provide a fence around your tower to prevent it from being an attractive nuisance to the challenge of junior climbers, or install metal plates covering the bottom rungs of the tower. Usually these can be on hinges such that when unlocked and opened, foot rails on the tower are exposed for climbing. The equipment cabinets can serve the same purpose in which case a ladder would be needed to get past the first fifteen feet or so to reach the foot holds if climbing is necessary.

FIG. 3.12.1 SINGLE POLE SITE WITH BUILDING

In another variation of the above project, a building becomes the main focus with a telephone pole used to support the antennas. One advantage of using a telephone pole is that it rarely requires anything but tacit approval by zoning boards. The pole serves triple duty. It is used to bring power and telephone service into the building, as a mount for an outside *dusk to dawn* light and. . . a "few" antennas. Figure 3.12.1 shows an operating system of this type. The basic assumption here is that the ground level for this system is till high enough that tower height and *tree-line* considerations are negligible.

3.13 An Intermediate Antenna Farm

My friend Bob, has no technical background whatsoever. He retired from a large food company and was looking for an interesting sideline which promised a secure income while allowing him sufficient freedom for frequent trips to his Florida retreat. Bob owned a few acres of undeveloped land on a nine-hundred foot (900' AMSL) mountain overlooking a capital city with a population of better than 300 thousand people. The height above average terrain is about 600 feet. Referring to the discussion earlier in this Chapter, that would be equivalent to a six-hundred foot (600') tower at street level! The plot of land is less than 5 miles from town. From the site you can actually see several nearby communities. Access is easy with a fully maintained and paved road. In addition, the site had plenty of AC power from mains and is served by a local exchange telephone company.

Easy to construct and operate, this intermediate system offers a good compromise between a basic system and a more sophisticated system with plenty of room to grow. There are both advantages and disadvantages to this arrangement. The construction advantage is based upon the premise that *the site ground is already higher than the surrounding terrain* and that the construction of an expensive tower might not "add" much to profitability. So, why build it initially? But some technicians dislike working with wooden poles as they are perceived as being hard to climb and work on. On the other hand, antenna hardware supply companies make plenty of equipment for pole mounting systems. And there are lots of poles in this world with skilled climbers, and other equipment (like bucket-trucks) to make these systems acceptable as antenna mounts.

Bob decided that his property would make an excellent antenna site. Further up on the mountain resides several high power television stations and a few public service and costlier private site rental facilities. Bob's business plan was to provide a lower cost more convenient facility for medium and low end utilities, services and communication companies.

He started by constructing a block building approximately 10' by 20' with a sloping roof. The building is surrounded by chain-link fence with a gate as seen in Figure 3.13. Wrapped strategically around the building are 6 husky telephone poles with foot holds nailed in each starting at about the 20' foot level. Bolted between a few of the poles are 10' foot sections of *Rohn* 45G tower allowing additional mounting opportunities. All pole mounting hardware is drilled completely through the pole and bolted on the back side.

FIG. 3.13
INTERMEDIATE ANTENNA SITE SERVING 13 TENANTS.

This system hosts an array of at least thirteen antennas and transmission lines which are bracketed down the side of each pole and enter the building through pre-installed PVC conduits 3" inches in diameter. Vents on each side of the building allow plenty of air flow through the building which is painted white to reflect the sun's energy.

Since the wooden poles and antennas are on the outside perimeter, any ice sheets fall not on the roof but to the outside of the poles. The highest pole is 60' feet from ground level (AGL). The poles were installed using standard wooden pole mounting technology, i.e. drilling the ground hole first; setting the pole into the hole (which was heavily creosoted); and finally positioned and back filled. The 2 wooden poles in front of the building entrance face the population center in the valley while the rear poles face less populated areas to the East and West of the site.

A wide range of services are accommodated from this site. Several paging companies, an FM broadcaster, several community repeaters, a few top secret government operations(?), and a 220MHz. trunking system. The smaller satellite dishes provide downlinks for paging and a "big dish" is used by a broadcaster. One of the poles is used as the drop point for power, telephone and exterior lighting. Inside the building, flourescent tubes provide light. There are 20 grounded power outlets, each separately fused at 30 Amps (breakers) positioned around the outside wall. Large rocks were placed on the ground along the side and back of the building to discourage vehicles from using the area as a "parking" lot (for you know what). In addition, over 20 copper rods were driven into the hard earth around the building. They were then connected together, to the power ground, the building copper buss strap and the telephone ground. A grounded heavy copper wire runs up each pole. Note also that the *tree line* is much lower than the poles and as such represents a mature growth situation not likely to grow higher than the poles themselves

FIG. 3.13.1 BOB CHECKS OUT CABLE NUMBERS ON ALL EQUIPMENT

Some clients have opted to mount dishes from the ground or roof level to insure they are secure yet high enough to avoid snow cover. Unlike small tower sections, wooden poles do not "sway" (much) and thus provide a very stable mounting platform for less complex antenna systems. Inside the building, equipment is positioned along the outside walls near power outlets, telephone service and grounding straps. Mounted on the wall are standard RJ-11 telephone interconnect boxes, (not a phone company requirement anymore) but still a good idea if phone lines are used in connection with wireless systems. The building has been provided with a fifty (50) pair cable entering from the West side and terminating into a standard telco-block using gas-lightning protectors. Customers connect to their control lines from a punch down block provided for their access.

Bob insists that all systems be well labeled and marked. It is crucial for service personnel to be able to quickly identify subscriber systems and connection points. The practice is often ignored and will certainly cause problems as time goes on as new clients are added so pay attention to good documentation from the onset and insist that your tenants and service personnel do the same.

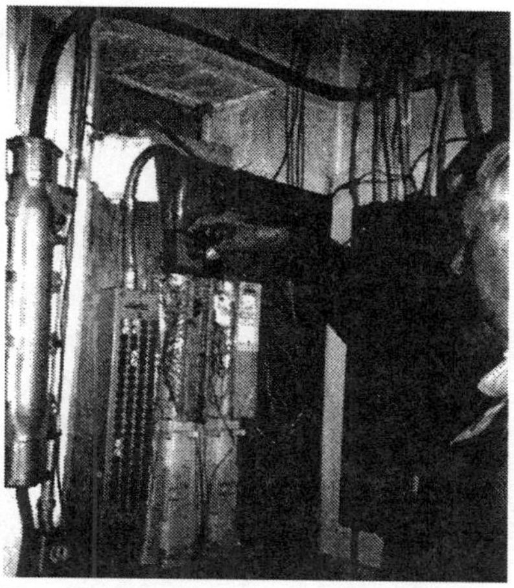

FIG. 3.13.2 TELCO PANELS USE GAS PROTECTORS

Bob also provides an active phone line at the site but this may or may not be advisable. It could be provided with toll restriction, i.e only 800 and local numbers can be processed, but with new inter LATA no longer requiring area codes and coded long distance carrier selection, the line could still be subject of some abuse. Should the courtesy be abused disconnect the line.

This antenna site is capable of serving about twenty (20) users without much modification. Assuming that each pays about $200 per month that is about $48,000 per year in gross revenue. Did I mention that Bob's real estate taxes were under $100?. A pretty good profit from a business that requires literally no attention other than depositing the monthly checks. What is an antenna site like this worth if you were to purchase it as an ongoing business? Some say about ten (10) times gross or $480,000. Realistically, however a site like this one, having sufficient land for improvements and expansions could easily sell for five or six times gross revenue. Not a bad return for under a $40,000 investment.

FIG. 3.14 TYPICAL SIDE MOUNTED PAGING ANTENNA ON AN ELEVATOR PENTHOUSE

3.14 A Rooftop Antenna Site

There is no question that a building or roof-top antenna system requires the least start up expense unless you run into serious structural problems or the present roof is in need of repair. With some simple planning, you can create a facility and market it for the best possible return on your investment which quite properly can be justified in short order.

It matters little whether the building is an office, apartment, factory or warehouse. Its position on the horizon and *view* from the roof is what is being sought by the telecommunication planners and this is what you must consider when preparing and marketing your building for an antenna site location.

Installing antennas are best accomplished by side mounting them to an existing elevator penthouse. This appendage is usually constructed of a steel reinforced frame covered with brick. But in the absence of these facilities, some manufactures can supply non penetrating *sleds* which can be positioned anywhere on the roof-top. They are steel frames upon which antennas may be mounted and are weighted down with blocks or metal anchors. The equipment room is the real problem. If you are too conservative your clients will not be thrilled about your proposal to have them put their systems in a janitor's closet. But on the other extreme I have actually seen situations where building operators assigned an entire top floor suite to the siting function. When calculated, the per square foot income from siting activity was clearly more profitable than normal rent or lease expectations.

This example of a single roof-top antenna system produces $400 per month for the building owner with only one national paging account.

CHAPTER FOUR

4.1 General Technical Considerations

The practical antenna siting examples presented in Chapter Three may not be ideal but they do have common elements which we will now explore in greater detail. Antenna sites tend to evolve from what is needed to what works and sometimes under very wide or even unusual sets of circumstances. Consider first the antenna mounting system. Then review the need to provide space (or a shelter) for the associated station equipment. Finally, there is a long list of indirect matters which need your attention such as twenty-four hour access, power, grounding, telephone lines, security and environmental considerations.

Site Savvy

> Antenna sites flourish both by adding new accounts and with the expansion of existing accounts. You must provide for both of these eventualities at your facility.

Your site will grow in two ways. You may conclude that your primary objective is to attract new business to your site but equally important is that your existing accounts may wish to expand their facilities. Taking this into account in the initial design will allow you to maximize site revenue by providing as many accommodating services as possible from the start. An understanding of the basic components of an antenna site which apply, more or less, to all of them is crucial to your success.

In this Chapter we will explore the physical plant and it's component parts. Discussed will be equipment housings, buildings and shelters along with the necessary interconnection facilities required. We will also overview the antenna mounting system and the importance of considering *and charging for* proper tower loading or how much stuff you can expect to put on your tower before it will fall over. Important to this discussion is the fact that you must plan for the future as you provide for present needs.

4.2 Equipment Housings

Obviously it is not enough to provide just a tower or a roof-top for your antenna site. You must also provide ground space, flat roof-top space or even a dedicated room to house your customer's communication equipment. And the *equipment* varies from company to company. For example, one ESMR communication firm requires a minimum of 300 sq. feet of fully enclosed, secure and environmentally controlled space. This could be a portion of a partitioned room in your existing building or more commonly a prefabricated equipment housing known as a *shelter*. Typically, shelters come in a full range of sizes, shapes and methods of construction from a wide variety of suppliers. Of immediate interest is whether you are required to provide the equipment housing or whether it will be supplied by your tenant. Obviously, you will want to analyze this aspect of your site very carefully. Large tenants will have no problem supplying their own shelter on your site. In fact some of them prefer to do so. With one single building installation they can provide total conformance to their engineering standards. In order to permit this at your site you are going to need a lot of ground space and a sympathetic zoning board who will not view your *land use plan* as one tower and one building.

FIG. 4.2
TYPICAL OUTSIDE EQUIPMENT HOUSINGS

FIG. 4.2.1
PAGING COMPANY "RACKS" SHARING A COMMON SHELTER

Typically smaller users like paging companies or repeater operators, do not prefer to provide their own shelter and would much rather just locate in your existing facility. This means that if you have a particularly attractive site to paging operators, you can expect to have at least several seven foot equipment cabinets lined up in a row all plugged into a separate outlets. These paging transmitters are most often all solid state. Smaller paging companies, transmit intermittently through the day with loading falling off in the evening. Larger paging companies providing wide-area or national service operate quite differently. They are usually high power (350 to 500 Watts) and transmission is virtually continuous ("key-down") all day and night. Put a few of these tenants in your building and the amount of heat generated will be more than sufficient to warm the building in the Winter. During Summer, however you'll have to provide adequate *air conditioning* to remove this heat or the transmitters will suffer *thermal shutdown*. What all this means is that your building will require adequate electrical service outlets, fusing, floor space, grounding and some pretty hefty air conditioning (2-4 Ton or 24,000-48,000 BTU). The maintenance and electrical service just to operate the air conditioners as a common building environmental service can be a formidable expense so do not overlook it. When a prospective communication company looks over your facility, a portion of their *evaluation sheet* describes the equipment placement possibilities. You can easily anticipate these needs by providing what is required. If not you will at least have to provide ground space for the positioning of their facilities. Typically ground space around the base of the tower should be no smaller than 50' x 50' or twenty-five hundred square feet. This amount of space could permit several shelters to be positioned around and near the tower structure. Use a 10' to 15' separation between structures. Crushed rock is used on the ground surrounding the shelters for drainage and weed control.

FIG. 4.3
TYPICAL ICE BRIDGE FOR CABLES

Shelters are usually set on concrete slabs, although several new designs permit direct placement on the ground or on footers poured under the support rails. There are certainly a wide range of shelter options and a few of them are listed to provide you with some ideas and sources for additional information to complete your design.

4.3 Typical Shelter Types

a. <u>Concrete Equipment Shelters.</u> When durability and security are of prime consideration solid high strength lightweight concrete shelters should be considered. This type of shelter can be supplied in a wide range of sizes such as 8', 10', and 12' widths; and from 8' to 36' feet in length as standard. The interior heights are roughly 8', 9' to 10'. When a single shelter is used for several paging services for example, the cabinet heights including cable exits are usually 7' in height. Because transmission cables usually emerge from the top panel of these cabinets, sufficient space between the cabinet and the ceiling should be provided. Accordingly, 9' or more is recommended. Some shelters actually come with rails mounted on the ceiling for easy routing of cables to the *portal* or the point in the wall where the transmission lines go out to the tower.

Usually all the cables exit the building at the same place rather than having the cables emerge at random points. A common *ice bridge* passes between the building and the tower base. Cables are mounted *under* this bridge so that slabs of falling ice from the tower legs won't damage or sheer the lines. In addition, all the cables can be grounded at the same point prior to entering the building. Cables should also be identified for easy location and tracing in the event an *individual* cable becomes damaged or is in need of replacement. Cable routing is an art and some serious problems can develop without consideration. If cables are merely routed out to the tower and strapped on a leg every 6' to 8' feet or so, until reaching the point where it connects to the base of the antenna; and the next tenant does the same thing etc. you'll wind up with a *large bundle* of cables all bunched together going up a tower leg. Not only is this situation hard on tower loading but if a bad cable happens to be in the middle of the *bundle* it is virtually impossible to replace because the other cables are attached to it. Identifying cables and where they go also becomes difficult. Managing the number and type of facilities installed on your tower is crucial. Even with a few users you will have many technicians working at your site over the years who know nothing of your site agreement with your client. They frequently install unauthorized equipment not a part of your contract. Accordingly you will need an easy method of inventory to discover any variances. Simply stated all you need do is to *count the number of transmission lines* and antennas and then correlate them with your site agreements.

Without control, a jumble of transmission lines and un-removed *dead* antennas can quickly accumulate. Loss of usable tower space from which revenue could be obtained is not the only problem. Antennas and transmission lines typically last on your tower for a relatively short period of time (about five years) before the effects of weathering, corrosion, vibration damage, ice, water and insect infiltration begin to take their toll. When it is determined that a system is degrading, service people generally install a new antenna and line very quickly to restore service. Often the old system is left in place *to be removed on another day*. If no one follows up on this removal, it can remain on the tower serving no purpose other than providing unneeded loading.

Concrete shelter designs provide the best protection from bullet, earthquake, and fire damage. Bullets? Unfortunately communication facilities are favorite targets for an irresponsible hunter or frustrated sportsman not just vandals. Power meter bases and shelters themselves are fired upon all too frequently. A single 22 caliber bullet fired at a transmission line at the sixty foot level on the tower can dent the cable sufficiently to permit continued service but at a reduced level of performance. Damage of this type is almost impossible to repair and the entire cable length must be replaced.

FIG 4.3.1
CONCRETE ROHN SHELTER READY FOR IMMEDIATE USE

In addition to the concrete design, many shelter types are steel reinforced, and insulated with either ridged foam or polystyrene. Typical *R-values* are between 6 and 19 depending upon the final location of the facility. Usually the walls and ceilings are one piece and sealed with butyl rubber compounds eliminating the need for external caulking.

In most cases the entire shelter is constructed at the factory and shipped on a flat bed truck and *dropped* into position using a crane. Attachment rings are an integrated part of the wall construction and are positioned to balance the lifting load. In addition to the shelter, electrical systems may be supplied as options and include telephone panels, custom electrical service breaker and distribution boxes, out side connections for lighting and generator attachment to internal equipment. Check with *Rohn Shelter Facilities®*, 1100 Industrial Road Bessemer AL 35021 for a wide range of shelter designs and options. **www.rohnnet.com.**

b. Fiberglass Shelters. A less expensive alternative to concrete construction are the fiberglass shelter types. Basically construction begins with 2x4 inch studs covered on the outside with a colored or textured fiberglass layer. Inside walls are either white vinyl or aluminum. Since the walls are standard 3 ½" thickness and 16 inch-on-center, fiberglass insulation can be inserted with an *R-value* of up to 11. The roof is of similar construction and can be insulated to as high as R-19. In addition various mounting arrangements can also be supplied at significantly reduced cost over concrete. These units are considerably lighter, less expensive and are easier to transport to difficult or rugged mountain top locations.

c. Prefabricated Shelters. For the lack of a better term another offering by *Enviro-Buildings, Inc.* **www.envirobldgs.com** provides a unique solution to roof-top equipment housings as well as standard tower site applications. This design consists of preconstructed light weight *panels* which fasten together using a *steel cam-lock fastener* ® in any number of different custom configurations. One application described by their sales people is the possibility of bringing the various wall sections up in an elevator and assembling the shelter (on skids) right on the roof. Typical sizes range from 10' x 10' to 10' x 30' shells, which can be equipped with *custom cutouts* for doors, air conditioning units and generator vents. This building technique also allows expansion as well as one large building with internal dividing walls and separate doors. The product offers some interesting possibilities for meeting local building codes and land use requirements too. A one building zoning restriction could start out small and keep expanding in size as customer requirements progress. These units should be used in more secure areas however.

d. <u>Cinder Block Buildings</u>. In Chapter Three a block housing was described. More like a shed than a shelter, a block building is inexpensive and can be designed easily to take advantage of the configuration around the base of the tower. Often a building permit is required along with an approved design. However, whether the shed is constructed of block, wood and siding or corrugated metal sheets; or is even a pole building, it should be fairly well insulated, secure from insects and vandalism. Typically a footer or steel *rebar* reenforced concrete slab is poured and the building constructed on top of it. Roof construction may either be a sloping flat roof to allow water run off or any other suitable design. Some thought should be given to the strength of the cross supports (rafters) as any ice slabs falling from a nearby tower could easily cut through ½" CDX plywood if the rafters are greater than 16" on center. The basic design generally is long and narrow, i.e. a 10' wide shed can be 30' feet long, with several doors and internal petitions creating rentable compartments and common use areas. Be sure to allow some secure space for your own needs to store tools, spare parts and landscaping equipment. Unfortunately with any other type of construction space is at a premium and storage facilities for site maintenance is the first expendable area to be eliminated.

e. <u>Miscellaneous shelters, mobile homes, trailers, truck bodies, wooden garden storage shelters, outside cabinets or anything else.</u> Obviously there is a host of other possibilities while you are waiting for your tower site to develop some *power tenants*. These are but a few of the possibilities which can serve in a pinch to provide facility housings. But your paging client is not going to be too enthusiastic about placing a seven foot cabinet with a retail value over $16,000 in that chinchy aluminum shed you got on sale from the local garden center. Even large *T-111* wood sheds, although a bit better, are not desired because of the ease with which they can be broken into, burn down, or more importantly actually support the weight of several transmitting cabinets. In fact this is a real problem with most of the units in this classification. Construction type trailers give the appearance of being a good solution to a temporary facility however their plywood flooring and under support frame will buckle under two or three typical base station cabinets. Smaller and less expensive shelters using plywood floors can be fitted with equipment racks from floor to ceiling mounted directly upon support joists. Communication equipment is then mounted on the racks but it is out in the open. This works fine if there is only one tenant having access to the building. Shared or common space however requires a locked (heavy steel) cabinet not only because it is preferred by the communication company to deter unauthorized tampering with their equipment but because, believe it or not, it is a FCC regulation.

One advantage noted of using a small trailer or truck body however is that should a municipality have some sort of zoning hoop when requesting a permanent shelter or building, often a portable housing on wheels will enable the service to begin operation while more suitable accommodations are being processed. This situation of course opens up a whole new can of worms which we will address in a later Chapter.

f. Roof -Top or Top Floor Storage Space. The final consideration in providing suitable storage space for transmission facilities on a roof-top, building or existing structure is that it is usually an afterthought. If we can get property managers, building superintendents or building owners to think in terms of *revenue per square foot* the decision process is not so difficult to comprehend. Roof-tops can be very valuable to communication companies because they are usually nestled in a high density (use) population center, have easy access to power and telephone lines and represent a relatively speedy way to get on the air. When negotiating a roof-top lease however a host of problems can surface which are not immediately apparent.

A communication company Site Acquisition Specialist will survey a given building's position within the *search ring* and if found to be a possible site, contact the building manager. If a *siting policy* has never been considered often the building management or owner creates most of the obstacles. Listening carefully, the building supervisor nods his head but has no authority to support or reject an offer. The story must go up the ladder. A typical first reaction from an owner is: *"I don't want anyone poking holes in my roof"*; or from the management firm you might hear:

"we have enough to worry about, it's not worth my time" etc. The communication company will propose some rent amount to the unsuspecting building owner which may or may not be realistic or a good reflection of what the site is really worth. If a deal is struck, the communication company will do it's best to design mounting arrangements to the exclusion of any other site tenant. In addition, the site rental agreement may contain an unsuspecting *walk-away clause* that favors the changing needs of the communication company (or the possibility of finding a better deal later some place else). These are negotiation issues which we will explore in detail but eventually it all comes down to *"ok now where can we put our equipment?"*

The usual response might be in the elevator control room (which incidently may not be *code* compliant in some states); or perhaps in a storage room, boiler room, maintenance closet or some other small ventilator access area. What your communication company wants to hear is that you have partitioned off a small office or dedicated an entire apartment to the function of housing transmission equipment. If the room has the same provisions as a ground mounted shelter, i.e. loop ground, power distribution, air conditioning and easy access to the roof (through transmission line entry panels) plus house lighting, a rest room and a work area for service personnel you'll have no problem writing up deals (and more than one deal is stressed) that will produce some big returns. The room need not be decorated. Carpeting or other amenities are not required. And once again consideration should be given to providing easy access to power panels in the basement or lower levels, telephone lines and other conduit passage ways which can be conveniently routed to the room. In a pinch rather than *core drill* 22 floors to bring up cables, a cable channel could be installed on an outside side wall of the building and disguised to look like a part of the actual construction .

When pricing your site, remember that there are two components. The first is the mounting for just the *antenna array*. Second is the *housing* for all the equipment. These two components are also influenced by a third consideration, i.e. the initial installation and where the installation might evolve in a few years. It is important to be specific in the description of the actual facilities provided at the time of the initial installation. In this way what is priced today can be augmented by additional rents for expanded facilities. Often communication carriers will want to write this into the contract as expansions. For example. If a company says they would like to start with *one box* and *three stick antennas* on your roof at a given price then at a later date move to *a six panel antenna system* and *two additional boxes* (for a total of three) in the future, they will want a price for the expanded system now. Obviously they won't want to pay for the expansion until it is implemented. Now things get interesting. What should you do? Price the expansion and write it into the contract? Or wait until the expansion is desired and then renegotiate? The problem here is that in effect you are giving the communication company an *option* to use your roof-top at some later date for more facilities. But you are doing so to the possible exclusion of another communication carrier who might also want to use you facilities now or in the near future. The question is do you want to turn away business on a inference of expansion or do you want to provide for as many facilities as is possible while systems are being constructed? The solution lies in a facilitative design from the start.

In some remote cases, communication firms will request that shelters (housings or cabinets) be placed directly on the roof. The technology is available for doing this both from a structural and practical point of view so the thought should not be met with blind opposition. In fact the engineering to prove roof-top mounting feasibility is rather expensive and you may sit in judgement of a better than average engineering specification that loading, roof integrity and access requirements will be met in every sense. And of course there is always indemnification and complete acceptance of liability for damages which should and as a matter of practice be a standard part of your site contract. If you have a site which serves a need for one carrier it will more than likely wind up being desired by others.

A true story. While performing site acquisition for a national paging firm I came upon an industrial complex of 30 or so low profile buildings, a few corporate headquarters, some office condos and all within a radius of about a mile. In addition the site was close to a five mile stretch of several well traveled roads.

In the center of this complex is an aging eleven story assisted living facility operated by an absentee owner, managed by what appeared to be an inexperienced young administrator. When it was determined that this facility would indeed serve my communication needs for the expanding complex, proposals were made. All were categorically rejected by the owner after inordinate mail and phone call delays. In addition to some of the reasons previously outlined there was one I had not heard before, "*Our CATV antenna on the roof might be needed again if the commercial TV cable went out*". Naturally the only course was to move on and utilize a back up site also under consideration.

About a year passed with my paging firm and a few others installing *fill-in* transmitters, two cellular company installs, a microwave dish and a PCS operator all building in the area nearby. Everyone was comfortable with their systems and had *moved on* to other construction projects. Out of the blue a call was received from the administrator of the original building informing me that the *owner* had "a change of heart" and wanted to explore telecommunications "siting opportunities" with me.

When the number of facilities installed in the area and an estimate of site revenues was totaled the lost opportunity to this building owner was staggering. It came close to $2.2 Million over the life of the various contracts. Had a top floor apartment been converted to an equipment housing there would have been lost tenant rent, about $150,000 over the same period. Accordingly the net potential revenue to the owner would have been about $2 Million for the same period. And guess who uses paging units and cellular phones to manage their building? Opportunity presents itself for a short period of time in the siting business and accordingly one must act quickly to take advantage of it.

4.4 Required Facilities

There are common elements for each antenna site regardless of the site's location. Some are obvious. Others require some thought and perhaps a little creativity.

Overall Size - Space must be considered and should always be the *largest floor plan* you can possibly manage, cost and ground space not withstanding.

Door Sizes - Door sizes should be no less than 3' 6" x 7' (or larger), solid steel and frame with good quality dead bolt and commercial lock sets. Weather-stripping and drip caps are desired to keep water from the door. Locate *down wind* so blowing rain won't enter the locks and freeze or corrode them. Make sure there is a tight seal all the way around to keep out insects.

Electrical - Considerations are metered current from mains, breakers, surge arresters, a good earth ground, lighting, and emergency back-up power. Provide at least four (4) meter bases with a total of 400 amp service, 60Hz. at 220 VAC single phase. In addition, consideration should be given to the addition of *transfer switches* so that critical use communication firms may substitute their own portable generator for power from mains in an emergency. Some thought should be given to whether you are going to provide power to your tenant or require that the tenant supply their own power. In some cases it may not be possible to provide separately metered service but this does not preclude the use of a meter to measure the power which is consumed. Standard meters may be purchased from electrical supply houses and placed *in-line* with the customer's equipment. The disadvantage is that they must be read routinely (quarterly), prorated and inserted into the monthly invoice which has the effect of making your monthly rental payment vary. It's always best to require your client to meter their own power rather than to include it in your rent invoice. It keeps your rent payment lower and there can be no costly shenanigans going on behind your back.

FIG 4.4 FOUR METER BASES

If you do supply power, you must be careful to accurately predict the correct power consumption of the equipment installed and to check these readings frequently. The reason is that peripheral and environmental ancillary systems such as cabinet ventilation, air conditioning or test equipment may not have been included in the information from which you made your calculations. Unfortunately communication systems don't come with consumer stickers attesting to predicted annual use. It is not enough to know that a given PCS base station uses say 3.2 KWH if another kilowatt hour is also consumed in the GPS downlink receiver, crystal oven, the modem or other alarm equipment that nobody thought to inform you about. These systems do add up and you should have no qualms in locking down the *who pays for the power* question. Once installed it is highly unlikely you'll have any access to what's contained inside their box which is consuming *your* electricity.

One final note. You can use a *clamp-on* current meter available from any electrical supply house for about $100. This device will read the approximate current flowing through a given line. If you don't have one that calculates the wattage you will have to do some math to determine the power in Watts, the Watts-per-hour in order to determine the KWH cost (1000 WATTS per hour) times your local power company rate. An example. A clamp on meter shows 4.8 AMPS at 220 VAC flowing through the cable to one of your customer's boxes. If you take 4.8 times 220 this is roughly 1056 Watts. Assuming this power is constant (and it may not be as you could be measuring it at a time when the transmitter is not on the air) take this number times 720 hours in a month for a result of 760.3 KWH.

Now take this number times your KWH electric charge. Don't forget to include surcharges and taxes. Let us say your KWH charge works out to 17.6 cents per kilowatt hour for the example. The approximate electrical cost will be $133 dollars per month. If this varies at any given time by 10%, we are talking a range of $1436 to $1755 a year in electrical expense. The lesson we must learn from this example is to *insist* that your customer pay their own power bill. Otherwise, be darn sure what the equipment draws before you set a rental rate.

From the example I don't mean to say that consumption rates aren't predictable. In many cases they are. If a 1 Watt (RF) FM radio translator box, about the size of a CD player, is pugged into your wall outlet (110 VAC) and draws 19 watts-per- hour, is on continuously with no variations in consumption, then at the rate specified above the monthly power consumption would only be $2.41 per month. In this case there is no need to have this tenant meter their own power. The solution lies somewhere between the two examples and you must decide. Finally, don't forget to allow yourself an escalation clause in your agreement to make adjustments for power factor, taxes, surcharges or price increases (or decreases?)

Another consideration is *the common area power consumption.* Another example. A paging transmitter might use a calculated $87.00 per month in electricity. At the same time however it may also generate a fair amount of heat such that your *common use,* 4-ton air conditioner must remove *that much more* heat from the room. If not considered carefully, you will immediately see upon adding this new communication tenant an increase in *your* electric bill. This can not only come from the increased heat that must be dealt with but from other abuses such as a technician forgetting to turn your building lights off or if a space heater is suppled for Winter building comfort, failure to turn these systems down or off. Don't depend upon signs to remind technicians working at your site to return systems to their ready state. Provide automatic lighting shut down features (a timer) and other techniques such as remote control monitoring. If your site is difficult to reach due to ice or snow in the Wintertime, a space heater left on inadvertently can eat up a lot of your profit if you can not return routinely to check on things.

Batteries - Some paging services, low power repeaters, microwave links, or translators may use several banks of *deep discharge batteries* to either power their systems all the time or as back up should the power be interrupted. A distinction should be made early on as to whether the system is powered *all the time* by the batteries with charging units replenishing the batteries continuously; or whether the system is powered by an *ac to dc* power supply with *trickle* chargers maintaining the batteries *until needed* during a power outage. The two systems consume different levels of power. Battery banks under constant charge can also produce explosive hydrogen gas or cause acid seepage which can eat right through a wooden floor if not properly maintained. Of particular concern with these systems is adequate grounding and venting such that even a minor lightning discharge might actually cause an unprotected battery to explode spraying lead and acid solution all over your equipment room causing a real mess. Venting hydrogen gas for example during charging is required by most code authorities and just makes good sense. Batteries must be housed in an acid proof container in the event the battery housing cracks or ruptures. And never set batteries directly on the floor. Have an approved eye wash water bottle (non-freezing) installed at any location containing batteries. It is a responsible action regardless of OSHA's requirements. In some instances the batteries themselves can take up as much room as an equipment rack. So don't forget that it's your housing space that is at a premium and you should be compensated for floor space used in addition to the equipment in your rental agreement.

Generator- Who supplies the generator? If there are a number of clients using a fair amount of power from mains and who are separately metered the introduction of a house (yours) generator becomes a complex problem. Nevertheless any remote site which provides an automatic cut over to a 15 to 50 KW diesel or propane generator is very desirable. Down-time means big bucks to communication providers and during power outages these systems actually increase in use so reliability must be provided with your site services or without them. However these systems can cost tens of thousands to install in addition to being capable of engaging flawlessly and reliably. If you do provide generator back-up be sure to provide immunity from liability for both equipment damaged by your generator going haywire or not kicking on at all. We live in a imperfect world and even though ninety-nine times out of a hundred everything will work just fine, a bird might just build a nest in the exhaust pipe and your generator will sputter and die when it is needed the most. Weekly testing and a regular maintenance contract is well worth the extra expense especially if you distribute the cost as a value added service to your clients. Make it optional so that if they don't need your generator they won't have to pay for it. Finally, be sure to place your diesel fuel or propane tanks behind a protective block wall. They too are favorite targets of our mystery shooters. A single hole in a diesel fuel tank can cause an oil spill of major environmental proportions not to mention an explosive hazard if out in the open. Remember that you are justified in amortizing the acquisition, maintenance and operational costs over several years of useful life by apportioning rent payments to include these augmentations.

There are some really good companies that make these systems function very well so take your time in soliciting bids for systems to consider. This job is best left to the expertise of the generator people unless you are confident of your electrical and engine skills. But it is well worth the effort in the long run. It may also even be required if you are responsible for tower lights or have a de-icing heating system installed on your tower. And oh yes, the power from mains should always be routed to your shelter *under ground* for the last fifty (50') feet because should an ice sheer fall on an overhead line it will most assuredly cut your power and create a potential hazzard to visiting service people.

Site Savvy

> Electrical service should be routed to the shelter underground to avoid any disruption of service or shock hazzard caused by ice falling off your tower.

Solar Power- One final challenge is what to do if there is no electrical service at the site. Power companies may be reluctant in their Tarriffs to provide service to a location if the cost is prohibitive or if there are large number of *servient tenants* to deal with in a right-of-way situation. Running poles across other people's land is not an easy process especially for only one end user. Delays can be incredible.

Some thought must be given to the more exotic power sources such as solar energy for truly remote communication systems. Essentially solar systems are really battery powered systems. The batteries are charged by solar energy. If daylight is insufficient to keep up with the energy consumption of the equipment, the batteries may need to be charged by a backup generator as well. Things can get pretty complicated here. A check with **www.sunelco.com/home.html** can fill you in on just about everything you need to know to get started in the area of solar power. But there are some annoying variables which must be considered. Typically sunshine is pretty short on a clear day in December as opposed to a clear day in June (in the Northern hemisphere of course) and the number of sunny days per year is also of major concern. The low power FM translator, in the previous example, running at 19 watts could easily be powered using this system. But most bets are off for the 500 Watt paging transmitter. Check out **www.huttoncom.com** for one manufacturer's approach to the concept and there are a few others as well.

Other methods such as wind generators may be considered as well but for operating a sizable facility we're going to need some good old all American electricity from the receptacle on the wall!

Lightning Protection - Did you ever see an installation where the line cord plugged into the wall and going over to the cabinet has a *knot* tied in it every eighteen inches or so? Well it seems that some radio engineers are convinced that if lightning strikes, even nearby, large pulses of energy might be induced into nearby telephone and power lines. When a power or telephone lead has several *knots* tied into it, one effect is to introduce an AC resistance called inductance. Because most random pulses consist of high frequency current (as opposed to 60 Hz.) the *knot* forms a *choke* which opposes the full impact of the electrical current flow actually providing effective protection against induced pulses.

As simple as this technique sounds, in order to protect solid state equipment, we are going to need a lot more protection from the top of the tower to the rods driven into the ground around the shelter. Grounding systems are also required by local codes, the power company, the telephone company and your tenant. Equipment suppliers, in order to validate warrantees and other design obligations, want proper grounding as well or they won't replace equipment if damaged. We will need adequate grounding for our system for other factors than just lightning! But the main reason is still lightning strike dissipation.

When thunderstorms arise air rapidly moves from one level and temperature to another creating static electricity which builds up tremendous potential differences. Soon the voltages get so high that electrons in the air (air is normally a stable insulator) become displaced creating a cloud of air containing free electrons all scattered about. This kind of air in this state forms a conducting *plasma* which for a time becomes a conductor.

What we have here is very similar to Neon when it is excited by high frequency voltage. When ionized and turned into a plasma, Neon, contained in an airtight tube, can conduct electricity and give off an intense light. When natural air becomes ionized it too can become a plasma conductor. It may only occur for a brief period of time but sufficiently long enough to allow the potential difference between the negative charges in the upper air to draw down to the positive charges coming upward from the ground. The *point* at which a path is established can be the ground itself, a tree, a tall building or *your* antenna tower. Once a *streamer* connecting the two charges come together a powerful electric current begins to flow. Positive ions travel up to the negatively charged cloud at speeds in excess of 60,000 miles per second (one third the speed of light.) This is the actual part of the strike which can be seen. When the path has been established it can actually repeat its flow again and again *over the exact same path* giving the impression that the strike is *flickering* a phenomenon that photographers just love to shoot. The positive and negative charges built up by multi-cell storms along a squall line or rapid movement of humid-dry air boundaries eventually neutralize each other and the danger passes. If a bright bolt displaces air fast enough, it can do so faster than the speed of sound resulting in thunder.

Another type of lightning is caused by storm clouds speeding by your position whereby reciprocal charged particles *do not image along the ground* but remain in the air. Potential differences continue to build until an *arc over* occurs through the ionized air. Now we have a path which goes from cloud to cloud. This type of lighting is sometimes referred to as *heat* lighting which discharges totally in the atmosphere. Should your tower be situated at the exact spot where the positive charges on the ground find the *stepped leader* (the zig-zagged pre-strike path to conduction) of the ionized cloud in the sky. . . **kaboom!** A great deal of the energy in the lighting bolt will pass through the tower into the ground and most of your other systems as well.

Obviously the anatomy of lightning is a science of its own and is well documented in excellent publications. For a more complete study of the subject you are encouraged to explore the phenomenon. In researching the subject for this brief description it was discovered that almost every article, book or paper had some little tip or idea which could be considered or incorporated into a working *antenna farm.* With the introduction of NOAA weather radar systems computer complexes, for example supply advanced warning of impending electrical storms. In addition to power surges and electrical spikes, management information systems or large computer operators are very concerned about momentary power outages. Such a power fluctuation or disruption can wreak costly havoc upon programs which are running at the time of the failure. So it is with transmission equipment. Pagers can be *cut off* in process, and calls *dropped* in mid-sentence with even a momentary fluctuation in power.

To meet this possibility large computer shops subscribe to weather services which use early warning radar systems to notify operators of an impending electrical storm *before* it arrives in the area. Once warned, the computer shop starts up a generator and cuts all operations over to internal power until the threat passes. When power is transferred under *controlled conditions* the transition can be smooth and reliable. This technology is easily applicable to antenna sites even with locally installed detection equipment providing a very high margin of operational reliability.

There is no shortage of theory as to what works best at antenna sites. One theory holds that it is only necessary to provide such a low resistance path to earth that any lightening discharge will seek only the ground path to the exclusion of other nearby systems. This method is the concept of a *good ground*, not always easy to accomplish in sandy soil for example.

Another theory suggests that all that is needed to avoid a strike is to *prevent* the negatively charged (step leader) from finding the antenna site when it approaches the million volt discharge level. The technique in practice is simply to provide a single *sharp point* at the highest part of the tower. Some theorize that because the single point at the highest level on the tower is so small, the maximum discharge level through this conductor can only be a few hundred volts, no where near a million volts. As a result, lightning it is reasoned, will ignore the point and move to an easier target with greater conductivity, like a nearby tree. Well, maybe.

But for as many methods to prevent lightning damage there are products which must be considered which do indeed provide various degrees of protection. If only it were that simple. It is hardly enough to provide a low resistance grounding system for lighting because there are a host of other systems which also require a good connection to an earth ground for proper operation. Power surges and transients from the power company need to be detected quickly and dissipated, (hopefully not through your customer's expensive equipment). Devices which provide this protection are called surge arresters and are marketed by a host of excellent manufactures such as *A.C. Data©, Joslyn©, Northen Tech.© ACT Comm.©, Transtector©* and others. One type generally referred to as a MOV(Metal-Oxide Varistor) is placed across the AC mains at either 115 VAC to 250VAC. As soon as a voltage greater than the threshold (voltage by design) is detected the unit shorts out. These devices allow normal voltages to pass but shunt, almost instantaneously (very fast), voltage spikes in excess of a stated amount to ground. In the presence of very high voltages the MOV is usually *fried* in the process. Otherwise the spike is passed to ground before it can short out the MOV or *pop* a breaker. The purpose is to keep the transient voltage from the AC mains from doing harmful damage to the communication equipment.

Other devices such as silicon zener diodes, gas tube or carbon transient protectors are incorporated into solid state circuitry and telephone line circuits which take care of minor transients. But the really big spikes need some heavy duty consideration.

Yet another unique concept in *lightning management* on towers is the *Dissipation Array System©* or *DAS©* offered by *Lightning Eliminators & Consultants, Inc.* **http://www.lightningrliminators.com** This system looks like a huge umbrella mounted on the top of the tower which interacts with a complex set of atmospheric principles to set up a *cone of protection* around the base of the antenna. The principal of its operation is fascinating and can be studied in detail from the company's excellent web site. Of concern however with the use of this system is that the *DAS©* adds considerable loading to a tower and because it is placed at the top also occupies the most marketable portion of the facility. But nevertheless the device offers some pretty impressive protection for troublesome locations plagued with frequent lighting strikes.

Fig.4.4.1 Possible DAS © effect?

Everyone has a lightning theory and I suppose I am no exception. One thing I have discovered through the years is that in a community antenna environment, side mounted antenna arrays rarely get clobbered even when lighting does hit the tower. It's the guy at the top that takes the hit. But there is a price to pay for being side mounted which we'll discuss when we look at antenna mounting and pattern distortion. As a matter of conjecture, I wonder what effect top mounted *platforms,* commonly used by cellular and PCS antenna arrays, might have on the dissipation principle employed by the *DAS©*. See Figure 4.4.1 Although the design principles suggest otherwise, it wouldn't surprise me to learn that the impact of a lightning strike upon a platform may offer some unique protection not previously envisioned.

Static electricity - Static electricity is another demon that requires some consideration. When walking on a wool carpet with leather shoes and then touching a door knob, the effect is demonstrated. In a like manner fast moving dry air whipping past a tower (the unmistakable *growling* sound of the structure's resistance to the wind can be heard) can create very high levels of static electricity which must be shunted to ground. The static charge is sometimes sufficient to light up flourescent tubes in the shelter without electricity even being applied through the wall switch. Another problem associated with static charge is the increased change in receiver noise levels and system degradation as well as possible solid state device destruction. That is why computer manufacturers tell you to *ground* yourself before inserting a plug-in card into your computer. Static electricity charges can actually destroy the card in the process of installing it. It's also why the cards always come in those grey static resistant plastic bags. Although not immediately apparent to a site owner, static noise and its effects can degrade operation of the client's facilities.

This problem has been around a long time going back to the AM car radio days when whip antennae were (and still are) mounted vertically on the vehicle. As the automobile *whips* through the air, the car antenna creates a static electrical charge. When a critical level is reached, static voltage discharges causing a *crackling* sound in the speaker. A vertically mounted rigid antenna wire actually extracts electrons from the air. When voltage levels build up to a *break over* point, the static electricity discharges back into the air. It was discovered early on that if a small *ball* was placed on the top of the antenna mounted on a moving vehicle there was a significant decrease in sudden or threshold discharge potential virtually eliminating wide band static charges introduced into the receiver.

Although minimized by newer modulation methods, the noise amplitudes caused by static energy discharging still presents problems today but in a different way. The RF energy of a static charge competes with the intended reception of the desired signal which must be sufficient to over ride the noise even though it is never demodulated or heard in the speaker. Even vertically mounted cellular antennas still have that little *ball* on the end of the wire and its not just for keeping it from poking out your eye or keeping a telescoping whip antenna from ever being extended again when completely collapsed. Some car manufacturers have even mounted their mobile antennas at an angle to cut down on the wind drag and decrease the amount of static. However, without going into too much detail it is not desirable to mount the antenna too far into the horizontal plane as omnidirectional coverage will be forfeited. Reception would then be *bidirectional,* i.e. perpendicular to the sides of the whip and not desirable in a mobile environment. Good tower top design, grounding systems, and cable mounting techniques can minimize this static problem. But should your site have a large number of battery operated systems, a static charge problem can create the potential for some pretty serious consequences. Hydrogen gas is easily ignited by static electricity. Don't forget the lesson learned from the dirigible Von Hindenburg.

Site Assessments - Before any grounding system is considered it is important to have a ground conductivity test performed at the site. This test computes the ground or earth's capability of conducting electricity and varies all over the country. Typically, in the dessert Southwest silicon, an insulator, and dry soil conditions make ground conductivity poor as opposed to loamy moist soil in the Florida panhandle where ground water might actually be saline (salt water is a rather good conductor). Your site will more than likely be somewhere between these two extremes. If your site is located on a Precambrian granite mountain top, you might not even be able to force a copper rod into the ground. Accordingly, a wide range of conditions and situations can exist which will influence the conductivity and effectiveness of your grounding system.

The conductivity test can easily be performed simultaneously when a Phase I Screening Assessment is performed. So what is an Environmental Site Assessment (ESA)?

Whenever you are going to lease commercial real estate to someone else especially communication companies it is a good practice to provide an environmental site assessment of the parcel of land which is used for the antenna site. If in the past there was any contamination within the scope of the Comprehensive Environmental Response Compensation and Liability Act (CERCLA) it will fall upon you and your tenants to correct if such a study is not performed before the lease is signed. But recent policy is that most communication companies require the results of a "Phase I" study as an addendum and part of their lease anyway. They need this study in order to establish their "innocent landowner defense" in the event of any CERCLA liability.

Here is the issue. You purchase a piece of land that used to be a coal yard. Now it can be used as a great antenna site. But over the years a lot of hazardous substances or petroleum products were dumped or buried on the property. It will be your job as the new owner to clean things up. That could get pretty expensive especially if the really bad stuff is still lurking about such as PCB's or Trichloroethylene. Even mountain top land could be at risk especially if power was previously supplied to a site (or passes over it). A power utility pole might have been struck by lightning causing transformer leakage around the base of the pole. If PCB's are detected in this case, (PCB's used to be a rather good electrolytic material used in pole mounted step down transformers) it would be necessary to clean up this spill. Fortunately the power company will have to assist you in this. It's probably a good idea to get a Phase I ESA before you purchase the land.

A Phase I Study (American Society For Testing and Materials ASTM E 1527-94) meets traditional due diligence methods for any antenna site valued at over a million dollars where there is even a hint of any environmental concerns. This could be some old geezer living across the street who thought he remembered when a previous owner's oil tank sprung a leak (40 years ago) and killed all the trout in a nearby stream. It is important to meet the CERCLA's innocent land owner's defense. Should ***any*** indications be discovered the procedure is to move to the next level or Phase II. What is being looked for is asbestos, lead-based paint or lead products, underground storage tanks, (UST) , contaminated soil or ground water or actually any suspected situation which could cause any environmental problem. Certificated environmental specialists perform the surveys which usually cost several thousand dollars (be sure to shop around). If the property is really badly polluted it might even be in the State or Federal "Brownfield" (I wonder what "brown" really stands for?) data base. If it's this bad, your state's environmental agencies might be able to work with you in restoring it to safety and in some cases provide some pretty interesting funding and cooperative programs to assist you. What this means is that the property has been disavowed by its original owner but with some help from you, it might be converted to another *highest and best use,* an antenna site.

If the land has never been used nor developed and it is highly unlikely that any environmental problems have ever been possible even nearby, you might get away with just a Transaction Screen. (ASTM E 1528-93). This certification carries less weight than a Phase I study and usually involves the innovative due diligent method on sites valued at less than $1 million dollars. The price range here is about half that of a Phase I Study. In addition to what might be found on the land, it is important to also note adjacent facilities which might be in violation with the effect of spilling pollutants over to your land. And it works in the reverse. Could there be a nearby hospital, school, stream or nature sanctuary that could be harmed, influenced or even sensitive as an environmental receptor of any problem existing on your land? This whole process is a bit complicated so a professional geo-technical specialist, engineering or testing firm should be consulted. These resources can make the necessary determinations, and if problems are discovered assist in abatement or containment and even coordinate the work. In some instances indemnification is also a possible solution but a good reputable consulting engineer is the way to go for sure. With a little luck your site will come up clean.

Two other reasons you'll need your geo-technical consultant is to do a soils or base study analysis and a ground conductivity study. The first study starts with boring several holes into the ground at the precise locations where your tower legs or monopole base will be installed. What comes out of this study is the nature of the soil. Rock or stratigraphic characteristics that your tower manufacturer will need to know are presented on graphs in order to design the correct tower base. What you'll get besides a nice written report is the location of any *refusal,* i.e. where the drill hit rock and couldn't go down any further. You will also get a standard geological column which describes the nature, type and usually the geological period of the strata plus any other *unusual finds* concerning the location. You could define *unusual finds* as human remains or Indian artifacts in which case all bets are off for installing a tower for a while.

In technical terms your tower manufacturer will need a soils report to design the proper size and type of the tower base. Results contained in your study include *allowable bearing capacity*, plus *passive pressures* at each boring location, any *depth that water may be encountered*, or evidence of water table variations, *blows-per-feet* when a penetrometer is used, a description of the soil in geological terms at each level encountered along with the dry unit weight and the buoyant unit weight of the soil types if different; and a description of any observed problems which might interfere with normal construction. The number of borings is also dependant upon the type of tower which is going to be installed. Self supporting towers usually require a boring at each leg to about twenty-five feet (25') or slightly more depending upon the type of base which is installed. Usually three types of bases, *pier & pad*, *a drilled pier* and a *mat style* are used as standard and are dependant upon the soil types. If the soils are loose or come up weak in support of the required density, it may be necessary to drill down even further to determine if better strata exists or if bed rock can be found to anchor the base on a solid footing. If the self supporting tower legs are greater than six to eight feet (6' to 8') apart at the base, drilling a pier foundation is the least expensive way to go. Guyed towers require borings as well at the base and at each point where guy lines will be attached to anchors. If the land is relatively flat and uneventful, only the base and one guy anchor point might serve the study if both samples are similar. Should there be a difference between the geologies of the two samples, then all the points should be analyzed.

The *boring core* near the base of the tower is carefully removed from the borehole and transported back to the lab where it is subject to a series of resistance and conductance tests by applying an electrical current to portions of the strata. From these tests, *soil resistivity* figures can be estimated to assist you in determining what kind of grounding system will be the most effective at the specific site.

These three (3) studies, the Phase I, Geo-technical Report and the Soils Conductivity or Resistivity Study are a good starting place for the development of your site and will save a lot of work and worry as time goes on.

Grounding - Now that you are aware that there are other reasons than the power company's requirement for a good ground, let us take a closer look at what is involved. In typical home electrical service wiring, the power company actually runs two (2) grounds. In 110 VAC service, the *white* wire connects to the return path and the *bare* wire connects to the ground lug of a typical duplex wall socket. Back at the fuse box you can observe that both the *bare* wire and the *white* wire in standard 110 VAC duplex wiring are actually connected together. This allows for ground fault detection devices to protect consumers against accidental shock when working around sinks, bath tubes and other potentially dangerous shock hazzard areas. But a lot has changed with the wall plug in the last couple of decades. At one time sockets had equalized slots, now they are polarized; i.e. one bigger than the other. At one time there was no ground pin, now a pin is provided on sockets extension cords and most major appliances. At one time you couldn't tell the capacity of the wall plug by the way it looked. Now wall plugs are matched to the current they carry. These and thousands of other changes are reflected in the National Electrical Code which underlies both state and local electrical codes and standards. In addition to these standards are a host of specialized designs used specifically in the communication systems environment. Common to all of these systems is a direct and major tie in to an **earth ground.**

"Providing a low-impedance path to earth for high currents from lightening strikes or power surges is an important requirement. Maintaining good earth grounding is also necessary to ensure all non-current carrying metal parts are at ground or zero potential for personal safety. Grounding also provides a reliable path to earth for low currents from Radio Frequency Interference (RFI) or static that will build up on system enclosures and signal return paths"

The above justification for the installation of a good ground couldn't be better stated by Elizabeth Robertson of Lyncole XIT Grounding, **http://www.lyncole.com** It is a must addition for your research notes.

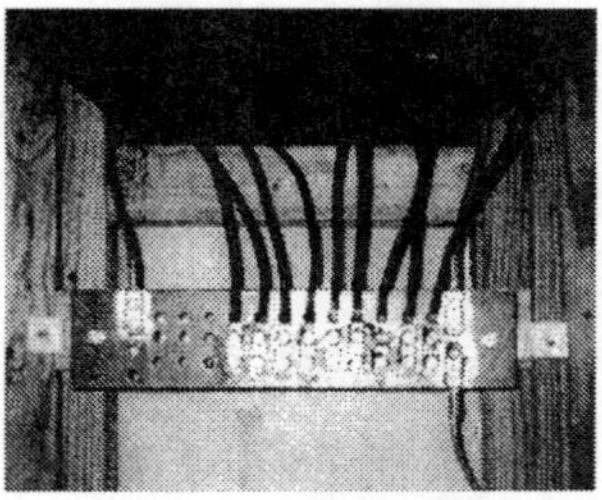

FIG. 4.4.2 EACH TRANSMISSION LINE IS GROUNDED TO THE MAIN GROUND BUSS

As with all electrical systems, and grounding is no exception, there are some technical standards (EIA/TIA-222-F) as well as instrumentation to assist in the determination of how the grounding system measures up. Most lattice tower, monopole, or wooden pole installations don't really make a good contact with the ground. Towers for example, are bolted to concrete pylons which incidently are fairly good insulators. Some AM broadcast towers, where the entire tower forms the radiating element, are situated directly upon an insulator intentionally. The lines which hold the tower in place are also insulated so the antenna will not come in contact with the ground via the guy wires. And the same holds true for the monopole. Therefore the tower structure must be attached by a low impedance bonding method to the earth ground. When a self-supporting tower is mounted on three or four isolated spread footings or drilled piers, each leg must be connected to a good pre installed grounding system. Essentially there are three types of grounding methods in common use.

a. <u>The Ring Ground.</u> The most common grounding system in use is the *ring ground* system. This system uses a large number of eight to twelve foot (8' to 12') copper clad steel or pure copper rods about one (1") inch in diameter and driven into the ground encircling the outer perimeter of the building, the waveguide bridge to and around the base of the antenna system and back to the origin. Although some techniques are suggested, each rod is no more than three or four (3' or 4') feet from any of the structures they encircle and are usually four to six (4' to 6') feet apart. The rods are then connected, rod to rod in sequence, by heavy duty copper

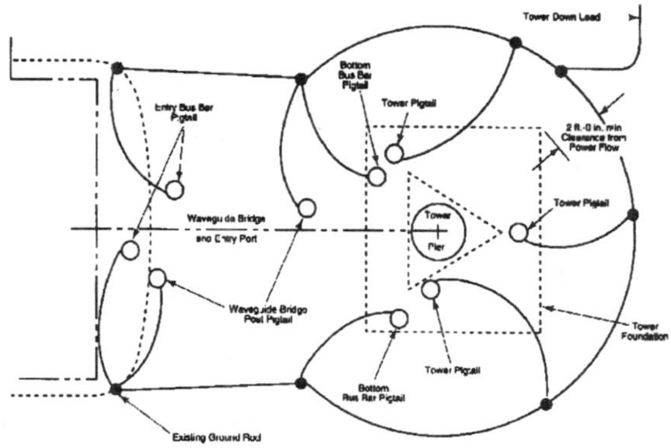

FIG. 4.4.3 GROUND ROD PATTERN FOR A RING GROUND SUGGESTED BY ANDREW

wire (4/0) to form a single *ring* around the entire complex. Key elements are then attached to the *ring* such as the waveguide support bridge (each end), the entry port buss bar (and connected to each transmission line shield) and finally to each tower leg. Connections are made directly to the nearest rod (not to the connecting loop wire). Most of the connections are welded joints except for attaching grounds to the transmission lines just before they enter the shelter. In this case compression nuts must be used as welding would destroy the cable insulation. See Figure 4.4.2.

Cadwelding© is a unique system of chemical bonding that joins grounding cables to most metals by first positioning a chemical putty like compound at the selected junction and then lighting a fuse with a match. When the compound ignites, an intense instantaneous heat bonds the connection together both electrically and physically. There is danger involved in using this process so follow the manufacturer's instructions to the letter.

b. <u>Chemical Ground.</u> This system is also sometimes referred to as *electrolytic* grounding or XIT© grounding. The system assumes that the conductivity of the earth around the facility is pretty bad. It is obvious this method could be used even if the soil is very conductive as the technology is effective in either case. Typically an auger hole is drilled vertically into the earth which is about eight inches (8") in diameter. The hole should be about twenty-two (22') feet deep for a twenty foot rod length. The bottom of the hole must be about half a foot deeper than the rod, and the diameter at the top should be about fourteen inches (14") to accommodate an access cover plate and preformed cap. This cap provides *breather holes* for continuous air circulation. The copper rod is about two inches (2') in diameter with a sealant tape at the bottom. This tape must be removed and saved before the rod is inserted into the hole. A compound known as Lynconite© and which comes in bags is mixed with water to a slurry consistency, i.e. like pancake batter, then used as backfill into the hole once the rod has been inserted. The cap is inserted over the top of the rod and the cover plate finishes the job. Typically the rod length should be no shorter than eight feet (8') but according to specifications could be as long as four hundred feet (400') deep. These compounds are not hazardous and typically one 20' rod replaces about 4 to 7 driven copper rods.

c. Radial Ground. This system is more than likely the granddaddy of all earth ground systems. It was born in the early days of AM broadcasting and generally consists of copper wires attached to the base of a tower, extending out from the center point in all directions around the compass rose. The length of each wire is slightly longer than the tower itself. The number of radials was thought to be governed by the principle that the more. . .the better. However with the advent of computer modeling it was soon discovered quite a few years later that after a certain point, any additional radials contribute little incremental improvement to either the radiation pattern or the effectiveness of the ground.

Arguments abound on this one issue of course but in general, forty to sixty radials, that's one every six (6) degrees around the compass, should do an adequate job. What is interesting about this ground system is that the ground itself, at AM frequencies is less dependant upon the earth. Sometimes this method is even referred to as an *artificial ground.* The theory is fascinating and easily tested today in a host of communications services such as Citizen's Band and other public service communications systems using an antenna called an elevated *ground plane*. By design, the *radial* wire depth is just a few feet below the surface of the Earth covered by dirt. At or near the base of the radiating antenna tower, at the focal point, the copper wires are often attached to *copper straps* about two to three inches wide. Because this ground system was also a part the antenna radiation pattern, multiple towers, in line, had their ground systems overlapping. Early AM broadcasting transmitters, and a few today, used huge amounts of power, in the 50KW range which generates its own "fire storm" of energy that literally got into everything close by. Power reduction regulations and changing listener habits have eliminated many of the "clear channel" broadcasters who used to provide reliable corridor coverage for thousands of miles at night.

Of even greater interest is that in some cases AM radio stations which can no longer be made profitable just go *off the air* for good leaving their tower arrays behind ready to fill a new function (which could be more profitable than operating the broadcast facility ever was). Many earlier towers were constructed of solid steel (rather than tubular pipes) and some are even self supporting. If these towers have been adequately maintained they are excellent candidates for antenna farms. One other advantage for sure is that the towers when originally constructed were probably out in rural areas. Because they have been there for some time the population center may have grown up around them providing a pre-existing use of the tower in locations which by today's standards would be hard to zone.

If you use an AM tower facility your grounding system may need some improvement because of its age and normal degeneration and corrosion. Any installation of shelters near the base of the tower will also need to take into account the location of the radial system. AM radio station studio facilities were rarely located near the base of the tower in high powered applications because the AM radio energy often contaminated the telephone system, audio and recording facilities. This is the reason why many communication companies refuse to use operating AM towers to mount cellular or PCS stations. It's not that they can't be made compatible, but it is a tricky science at best and can fall out of alignment at the drop of a hat. *Counterpoise AM broadcast arrays*, do offer a modern approach to collocation but unless the technology is appreciated by the Site Acquisition Specialist they are usually under walking orders to avoid AM radio towers. One final point. Should you acquire land near an AM radio station and erect a tower for communication systems, you might be surprised to learn that your tower will actually pick up RF energy from the AM Station. If it happens it could couple rf energy into your customer's audio systems. In addition your tower may also distort the pattern of the AM station if placed too close to the radiating tower. Fortunately this problem can be corrected by *de-tuning* your tower at the AM radio station frequency. Once *de-tuned* your tower becomes invisible to the AM radio pattern and stray RF picked up by it is shunted to ground allowing clean uncontaminated operation for your customers. The FCC regulates the proximity of your proposed tower to an AM radio station for the sake of both parties. Just because your tower is not effected, the tower might very well effect the AM radio station and thus must be rendered neutral at *your* expense. In general, it is safe to say that *he who is first is the one who must be protected*. We shall see how this same principle also applies to co-location when interfering systems are encountered at your site.

Site Environment - Previously the importance of good ventilation, heat management and air conditioning was described. In addition to these obvious environmental elements is the importance of providing self-sealing trash cans for technicians while servicing the site. Without them, it won't take long for the floors to build up with spent fuses, wires, insulation, small parts, packing materials etc. Of greater concern are soda cans, food wrappers and other organic waste products so delightful to ants, yellow jackets and the wee little mice who soon will discover that your shelter makes a fine home for their family. Mice can squirm their way through the smallest openings and enter your shelter to enjoy an environmentally controlled home. They like it so much that they will build their nests right in the equipment. Their *droppings*, which are not only unsanitary are painfully destructive to aluminum and copper. A term we refer to as *ratified* or *mousified* equipment strikes terror in the hearts of most technicians because anything short of total replacement means that the circuit board or chassis is a *time bomb* waiting to go bad in the middle of the coldest night of the year.

Wasps can also be a problem both in your shelter and on up the tower. Their mud nests are found in the *open ends of pipe*, in between transmission lines and in the corners of your shelter. They also can be pretty aggressive if disturbed on a warm day. Careful attention to this pest is a must because if a tower climber encounters a wasp nest while hanging on the tower there is no place to go. A good *drop'm* dead wasp spray is a must around the site and in your utility area.

A word about serpents. I don't think any shelter with warmth coming from it somewhere is immune from friendly and not so friendly snakes wishing to rest, molt, lay eggs or give live birth to their progeny. So look for evidence of snakes everywhere and carefully inspect transmission lines and other locations before you start moving things around. That piece of coax you want to move could be a black snake! And I've got to tell you, nothing scares the poop out of you worse than seeing a snake look you right in the eye, two feet away when you are not expecting it!

Finally, I once had a groundhog who built his borrow under a shelter. Since he didn't cause any problem I didn't think it was necessary to evict him because he wasn't paying any rent. He actually kept the weeds mowed down around the edges of the building. However I have heard that these fellows like to chew on wire now and then so keep an eye out for destructive evidence in your routine inspections. I'm sure our annual dowsing of weed controller (we don't use the word *defoliant* anymore) wasn't too helpful to his genetic code.

Site Access and Security - This subject is highly objective with lots of different approaches and solutions. If your site is in a remote location chances are there is a road leading up to it. If it is a shared road there may not be a gate at the entrance. Otherwise to keep unauthorized people from driving to your site to either *party* or vandalize your facilities, a gate is usually employed to *lock and block* the road from vehicle access. A good quality lock at this point deters most occasional trespassers from driving further. The gate, unfortunately is expensive and in most cases can either be obtained pre-made from an agricultural supply house, chain link fence company or actually made to order by a local welder.

FIG. 4.4.4 WELDED BOX HOUSES LOCK FOR PROTECTION

Some of the considerations when using a gate is that in addition to the installation mechanics, you must figure what to do in the Winter if you get three feet of snow which then hardens to ice. If the gate swings *in* you can't get a snow plow behind it without opening it. If it swings *out* you may have a set-back problem. A large heavy duty chain drawn between two large *bollard posts* solves this problem in Northern snow prone areas. But if the span is longer than twenty feet (20') or so the sag can be lifted for snowmobile access without unlocking. Preplanning and experience will eventually provide the best solution in your situation.

Another strategy which works fairly well but is less professional, is that access roads should not be paved but have *deep ridges* and *potholes* in them on the way up to the site. A nice paved road can be accessed by passenger cars but a *challenging* road will require a high axel four wheel drive vehicle which most technicians and site acquisition folks use as a matter of routine. Should foliage be permitted to grow out from the sides of the road, branches and briars are likely to scratch a passenger car's finish and thus serve as a deterrent. Naturally these techniques detract from the professional appearance of your site and can become a source of agitation to your clients who do not appreciate this method of forestallment.

Posting a *No Trespassing* sign is also important to establish the needed justification for prosecution should your security system be effective enough to catch someone in the act. Person's walking around your road gate also trespass but are less likely to cause harm to your property. If your access road is part of a right-of-way, be sure to have a solid agreement with the servient tenant in order to co-prosecute. Establishing *authorized* access can be equally as tenuous. You actually *want* site finders to walk around the site to scope things out. A good sign indicating permissible site access and phone numbers might assist in this objective.

But here is an example of what can and does happen frequently. A technician is dispatched to your site and finds he does not have the gate key after driving a long way. Usually they have the tools so they *cut the lock*, (unauthorized trespass?) access the property and leave. They don't tell anyone. Now you have an unlocked gate that you don't know about and you are out the price of a hefty lock. It happens all the time.

If you didn't record the lock type and number or have duplicate locks for the gate in stock, you will have to reissue many keys. I've seen as many as 110 keys typically re-issued in a similar situation. So when you do purchase your lock, purchase a dozen all *keyed* the same so you won't have to reissue keys. It is also a good idea to leave a spare gate lock at the shelter (unlocked of course) to be used if a technician finds the actual gate lock removed. There is always a technician who wasn't in the loop when you changed the key and the same situation will repeat itself. But technicians tend to want to leave everything normal when they leave, so help them do it even if you suspect they were the ones who broke the rules. If it's *in the family* things tend to resolve themselves. The opposite side of this issue however occurs when a technician is fired by one of your clients but retains the keys to your site. You rarely will be informed of this situation. An annual survey of authorized key holders is in your interest and provides responsible management of your site. If a discrepancy pops up negotiate the problem with your client's facility administrator and note the discrepancy in writing. The first unexplained infraction at your site involving an "insider" will produce a suspect that can be turned over to authorities. If technicians employed by your clients are aware (by site rules notices) that you control the situation in this way, they will usually not abuse their key possession and turn it in to avoid suspicion. Your clients will never question reasonable and efficient site security. It is part of what they expect and pay for.

Another popular technique is a *lock chain*. This method makes a chain out of the clasps on each lock, or connects a heavy duty series of chain links or rings, such that if any one (1) lock is opened the chain will open and access is permitted. The purpose of this technique is to allow your clients to use their *own locks & keys* (or combination locks). The problems associated with this method are obvious. After a time, mysterious locks appear and rusted ones never are removed. The chain gets larger and larger until someone puts one of those *89 cent specials* they got at the dollar store in the sequence. A good smack with a hammer and that one opens and then the chain is stolen and you are back to square one. The logistics of using this method is that it has to be opened the first time by you (or somebody) in order to insert the series lock. Not a bad method for a few clients but it definitely gets cumbersome over time.

A *lock-box* method works fairly well if the *lock box* can survive a sledgehammer strike and still function. Lock-boxes can be pretty hefty and bolted from the inside to the bollard and protected from all four sides by a welded steel plated *box* which only permits a hand *in* to work the combination. By using the correct combination, the box door opens up and a key on a chain *pops out* to open the gate. This would mean that technicians can still use a key if they have one but may use the combination on the lock box (or call in for it) if they do not.

An interesting variation of the combination type lockbox is gaining favor nationally. It is called the *Supra Advantage-tm*. **www.Supra-products.com**

FIG. 4.4.5 ACCESS CARD USED WITH *LENI* SEE TEXT

Unique to this system is that the key box or *lock box* is a "smart box". When it is used in conjunction with the *Supercard-tm,* the system allows a limited series of historical reports to be generated. The *Supercard-tm* must be validated monthly via a dial up computer program called *LENI* or Link to the Electronic Network of Information-tm. Used successfully by real estate sales people, the *Advantage* system has many products that can be utilized. A single *Supercard-tm* can access hundreds of sites all over the country. Access can be controlled fairly efficiently over time via computer update and custom coding. The only disadvantages are that the *Supercard-tm* is relatively expensive (about $100.00 each), is battery operated and must be used with memorized PIN and control numbers. The battery can be too low to operate the system and it is not a common battery type. Ample warning of a "low battery" state is supposedly given with regular use, but in my experience the unit just quits working.

A lock-box can be concealed or placed at the shelter so that the technician must first walk to the shelter, get the gate key then walk back to the gate. If the gate is at the bottom of the hill or it is far from the shelter these methods will be impractical. If the road gate is near the shelter then perhaps the electronic strike method or card access system could be considered. These systems are hard wired into the shelter security system. In some cases a gate with a *faux pad lock* works fairly well. This is a gate pin on a super chain with a huge lock, in plain sight with a big red No Trespass warning ("Violators will be shot... survivors will be shot again") but the whole system is unlocked.

One really neat idea is to place a single key with the nearest local police station. Usually police stations are open twenty-four hours and your technician can report to the station first. They "sign-in as an authorized user" (from a list) and take the key and proceed to the site. When they are finished, they return the key, "signing-out". If this service is a valid billable security service it has the added benefit of letting the officer on patrol know that access to your site is in progress. It is well worth the effort and expense and can be an efficient method for the phone company, power company and other routine service people using the site infrequently. Should your police department not be authorized to invoice for this service perhaps a service equivalent contribution to the FOP or PAL would be appropriate.

There are two reasons for blocking and locking your access road. The first is to prevent vandalism and robbery. Stolen base stations are usually shipped out of the country and find themselves in service elsewhere. I worked for a radio station a long time ago and upon coming in early in the morning to "sign-on" our 1KW broadcast transmitter was gone! Lucky we had the old stand-by in the back room. It was obvious what my first newscast was about. The FBI has jurisdiction over these matters and eventually traced the unit to South America. We never did get it back. They even took the spare *finals* (tubes) out of the stock closet. A professional job!

If your tower houses local police or county authorities it could be targeted by a disgruntled citizen. I'm sorry to say that one of the new breed of cordless hacksaws can do a job on a lock shackle, transmission line or even tower leg in short order with a few high quality Carborundum blades and one or two extra batteries. Your objective as a site owner is to prevent this from happening.

The other reason is to prevent unscrupulous home re-modelers and handymen from using your remote property as a dumping ground for environmentally hazardous substances. Asbestos pipe insulation, shingles, siding or floor tile, old tires, paint cans and other materials are what you'll find on your land primarily because these bad guys have no other place to discard the stuff legally without an excessive dumping fee or a permit. The problem is that if they get away with it, it becomes *your* problem! Fine notices for dumping can inform the public of the severe penalties for doing so. I'm not suggesting a token $300 for littering sign you see along the highway. Put the big buck fine on the sign, check with your local environmental offices for the Public Law, code and penalties. Then quote this data on your posting.

Now that we have addressed the road gate and are standing at the shelter we can look at a few additional options for site security. Under normal conditions the entire complex should be enclosed with a minimum of seven foot (7') chain link fence with three strands of either barbed wire or razor wire (really nasty stuff) at an *outward angle* from the shelter. Both the top and bottom of the fence should be supported with rails or pipes to prevent pulling the wire up from the ground. The gate should be large enough to back a small truck up to the shelter door to unload heavy equipment. Crushed rock at this point with a small gate makes loading facilities very difficult. A small narrow concrete walkway which will handle a hand truck is all that is needed. Once again the fence gate has similar problems associated with it as did our road lock. The only advantage is that it will be much closer to the shelter where a new more modern electronic strike lock or combination lock may be employed. If you are just starting out, simple heavy duty clasp locks and keys may be used. However you may want to consider an integrated total protection system with plenty of features. Such systems provide access control, power protection, climate control, tower light monitoring, with plenty of extra channels you can "rent" to your clients so that they can specifically monitor their own parameters or control their own unique functions. However most larger paging companies and PCS users have all their own environmental status monitors built in and don't need your systems. Smaller users may be very pleased to have a few control options provided by you at an additional fee. Although there are a wide range of suppliers for these specialty devices a good starting point is the TSM 2000 produced by Hark Systems **www.harksystems.com** My experience with this unit is that it is clearly described in their literature, is modular in function and priced so that you can configure your system as you need additional capacity. Typically, you should also consider control access logging, i.e. who came into your building and when; how long did they stay and when did they leave? What is the room temperature or were the lights left on? Generator testing and other functions can also be accomplished remotely including ease dropping into the site remotely or sending back slow scan video from other ancillary equipment installed.

Now, if someone calls to complain (it is always a complaint!) that they have no access card or key, will you be able to "buzz" them into the site, even if you are attending the IWCE conference in Las Vegas. If a monthly print-out of all the monitored systems at your site is required it can be accomplished by dialing up the system and dumping the data to a local computer. Perhaps the system should send a page or e-mail if the tower lights fail. A good basic system should be capable of accomplishing these functions as either standard or as options. It should also not be unreasonably priced. Some systems could be hot monitored at a security control location by a commercial service. Should these services require radio communication you might find that a service exchange is workable. Be creative.

Obviously common to any access system is a dedicated telephone line which should be installed at the site. If this is impractical, the use of a modified minimum call (bag-type) cellular phone works very well when operated from a 12-volt power supply. The circuit or unit should be protected to prevent anyone from from clipping "on" with a butt-set and calling Paris, France to check on the weather for the week-end (at your expense). On the other hand, it is also not uncommon to provide a local line for the convenience of your customers as well.

Finally if your antenna site is on a building there are similar concerns and some new ones. Most buildings have security systems installed, keyed entry and perhaps even authorized use of the elevator to certain floors. However a lot depends upon the type of building. Apartment complexes or office buildings usually won't require extraordinary access procedures but try to get to your roof-top antenna site when located on a casino or government building where security can be more complex. A good start here is the rental agreement itself permitting access and setting forth the procedures required to enter the complex. In most cases the access required by your clients will be at a minimum and thus will not be much of an issue during operations. However during construction and set-up the procedures become much more important and relations with building personnel can become strained. The best policy is to inform building personnel at every juncture in writing (or with a FAX) of your access needs and scheduling.

CHAPTER FIVE

5.1 Tower Systems and Mountings

Whether using a tower, a pole or a roof-top to operate an antenna site it is always necessary to plan ahead in order to maximize the number of facilities which can be placed at the location. Poor planning and positioning of actual antennas can cause pattern distortion, interference and many other physical problems which can put the tower at risk of falling over or compromise the physical integrity of a roof. Safety issues must also be addressed along with the possibility of creating harmful Radio Frequency Radiation (RFR) or at least the appearance of such. In this Chapter we outline tower types, mounting configurations and a few things to keep in mind when planning your facility from the start.

5.2 Tower Suppliers and Accessories

Steel is the a slang term for tower parts. Towers can be supplied from many reputable companies operating within the United States, Canada or even foreign countries. U.S. based dealerships and factory direct orders vary from company to company. Because towers and tower components are generally heavy, selecting a tower supplier nearby to your site can save substantial shipping costs. When a tower is ordered and shipped the transport company is usually not required to *off-load* the *steel* from the flat-bed trailer. Therefore a crane should be standing by to unload, as well as to erect, the tower sections or pieces. A crane rents by the day along with an experienced operator so it is a good idea to either contract for the entire installation or to understand the many factors involved and the sequence of events when managing your own construction. Project coordination adds about 15% to 20% to the cost of the tower installation. If you are experienced in construction or have a friend who can help you, the savings can be worth the effort. The downside is that if anything goes wrong on your job, such as an inexperienced worker or other catastrophe resulting in an accident etc., you will not be covered to the extent your supplier or contractor would have been either by liability protection or with replacement materials. Experienced *riggers* and *steeplejacks* do know their business. They waste little time in erecting towers. They have the tools and are actually amazing to watch. If you contemplate constructing more than one tower, the first project might best be served by contracting for the entire job and using it as a learning experience.

5.3 Using Local Subcontractors

Most tower suppliers who provide installation services, use *local subcontractors*. These subcontractors are usually approved by the manufacturer and must bid against other local or even regional subcontractors for jobs. If you plan to coordinate the construction project, you can get to know several subcontractors and their respective employees rather quickly. Obviously this is a good resource for your future construction needs. Tower manufacturers realize that tower construction and erection is a necessary part of their sales effort. Except for highly customized installations where tower manufacturers must provide direct warrantees for their product, as a general rule, they don't much care if you coordinate your own installation. All they want to do is sell the product. However they don't want the reputation of a failed installation either should you erect your tower incorrectly. Many large communication firms use turnkey installers which have been selected by competitive bidding. In other cases manufacturers set up *joint ventures* with installers or have their own subsidiaries to do the job for their clients. Then too, installation services could be a part of the overall order. This shouldn't be a major consideration if your business plan includes only a few towers. But it is good to know a little about how the process works from the beginning.

5.4 Seasonal Construction

Time of year for construction is also a factor effecting price and available crews. In the Winter months, special considerations for excavation and concrete pouring where there is a possibility of freezing must be taken into account. This may represent slightly higher foundation costs. However the incidence of work for local contractors taper's off and these companies might be looking for work to help meet their payrolls. Crews also move around a lot from job to job. It is not uncommon to find subcontractors using specialized foundation drilling equipment working in remote areas hundreds of miles from home installing foundation bases for towers that won't be needed for months. Once the equipment is in a given area it is a lot less expensive to do a series of foundations at one time. And one more thing, in many instances these crews work a seven day week in order to maximize their leased heavy construction equipment. The ideal negotiating strategy is to be in *no hurry* for the foundation allowing the subcontractor to work through his priorities and use your project as a fill-in job. However in the course of foundation installation be sure to "close the hole" as soon as possible to avoid safety hazzards or vandalism.

5.5 Professional Engineering Certification

In its informative little booklet entitled *Tower Specifications*, PiROD Inc. **http://www.pirod.com** a leading supplier of towers and accessories lists a few other concerns. Registered professional engineers (PE's) on staff can provide, along with your steel order, specifications that have been certificated in accordance with codes and standards. These so called *wet diagrams* are actually tower specifications which have been stamped or sealed with the PE's original signature. They are quite handy before local zoning boards. The specifications usually address a host of issues including structural integrity of the base design, the tower itself, suggestions on how many antennas and transmission lines can be installed on the tower and what to expect in your area of the country with respect to wind, hurricane, and ice storms. Having certificated diagrams demonstrates your due diligence also giving the local code inspector something to use as a point of reference when inspecting your foundation and approving your project along the way. Since the PE's actually work for the tower company, they are familiar with their product. Any failures or special circumstances that may have taken place with their customers in the past forms an experience base protecting your design. PE's can answer technical questions quickly and as a matter of course are experienced and can deal effectively with local code enforcers.

5.6 Ice and Wind Loading Considerations

Local zoning boards love to have PE's testify at their hearings but this in many circumstances is impractical and unnecessary when the design specifications for your tower have been pre-certified in accordance with established criteria. The fun part comes when the zoning board begins to understand some of the criteria and then attempts to set their own rules. An example. You may find that compliance with the current standard, EIA/TIA-222-E, for ice loading meets your insurance and construction needs. This standard provides two options concerning ice build up. The first option takes into account that an ice build up doesn't usually occur at the same time maximum winds can be expected. Thus ice loading at 75 percent of the required wind load, i.e. 87% percent of the wind velocity, is stated as being what the tower has been designed to survive under a recommended loading. Another approach is to require the maximum tower icing at the maximum wind velocity! It is obvious that if ice builds up on tower parts, considerable weight must be added to the structure. If the standard is *maxed out*, the zoning board can literally price your tower out of reach by establishing their own criteria. Suppose they direct that your tower must survive an 80 MPH mile wind while holding one inch (1") of ice build up. The cost to achieve this reliability is well within the tower manufacturer's capability. However these towers are not usually found in their catalog nor are they inexpensive. Some feel that it is the zoning board's way of saying: "... *we realize that the Federal Government has instructed us not to restrict your tower but they have given* ***us*** *the right to insure compliance with local codes and standards".* The effect here is to make your project's profitability more difficult. The alternative to restrictive coding is to insure against improbable loss. It is a lot less expensive to build the tower system to the national standards under EIA/TIA-222-E and insure the prospect of an improbable occurrence of ice and wind knocking down the facility at a later date.

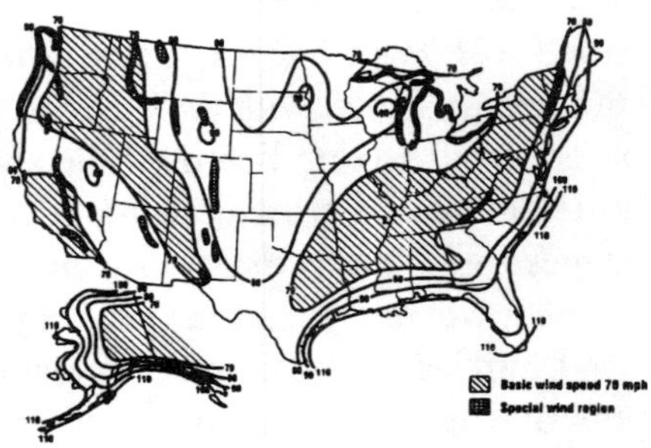

FIG. 5.6 WIND SPEED MAP SHOWING TENDENCY OF PREVAILING WINDS FOR DETERMINING TOWER LOADING.

Ice and wind are not the only hazzards to dropping a tower. Tornado, hurricane, plain old "rust" along with improper antenna installation can do the job very effectively. Most of the standards can be found in the *Electronics Industry Association Structural Standards for Steel Antenna Tower and Supporting Structures.* As the basis for design of minimum acceptable standards, the code forms the starting point in tower design. You can of course specify additional capacity to these standards as a margin of insurance against the unforseen or less likely occurrence of unusual conditions. But the tower design taking into account all the standards can be corrupted very easily when technicians start to *hang* their antennas and transmission lines without supervision. After all, your objective is to attract customers to the site. It is real easy to add five or ten antenna systems to a tower while at the same time departing from the original design criteria. And there are more codes to reference as well. Standard Building Codes, Uniform Building Codes, even the National Electric Code all reference communication installations. Another one is the American National Standards Institute or ANSI. That is why it's a good idea to have a PE sign the specifications. The certification gives your project the credibility it needs.

5.7 Fall Zones

A *fall zone* is another new standard that zoning boards are imposing. The presumption being that the ground around the base of the tower should be fenced in for a circumference which is equal to the height of the tower as the radius. The possibility of the entire tower falling over in a straight line from its base is remote. This regulation can add acres of land to your ground space requirement and multiply the chain link fence cost by several magnitudes. Even guyed towers, where three triangular guy line points support the tower, presents little justification for a 100% fall zone. If the anchor point is improperly installed, is too light, or is sabotaged there is a possibility that the tower will collapse along a plane perpendicular to the lost support line. However it is highly unlikely that all support guys will fail simultaneously and towers usually collapse in a distorted gyration of sequential steps resulting in a pile of rubble well within a 100% fall zone. Because there is little engineering justification for a 100% (or greater) fall zone, zoning boards may be inclined to distort the danger of this occurrence in order to discourage installations and evoke public concern. Self supporting towers tend to fold over on themselves and the only monopole failures reported were a result of an improperly designed base. Fall zones of 25% to 50% are well within reason. But this whole issue is a relatively new specification and has rarely been applied to power company poles, high voltage transmission towers, or even trees planted along Main Street USA or even in your own back yard. Trees tend to fall over with wind and ice as well, and can cause just as much havoc as a tower falling over. But a well designed tower, properly installed is a lot less likely to cause property damage or injury than that old oak tree in your back yard.

5.8 When a Tower Falls Over

When a *super* tower (800' or more) goes down, it can create a real mess and usually winds up on the front page of a lot of newspapers. The fact that the print media is competing with the broadcast media might have something to do with the coverage. Indeed, from time to time there have been construction accidents in which portions of a tower failed during erection, taking a worker with it. Thus in the minds of the regulators it seems to be pretty risky business so they want assurances that your tower will not wind up on the front page of the local newspaper.

Many of these new codes deal with the same issues from differing perspectives, i.e. one code deals with wind loading in terms of *psf* or wind pressure, while another refers to wind velocity expressed as *mph.* The engineering is the same but the definitions cause confusion and the appearance that the industry is unsure of what it should do. This fact has been the subject of thousands of transcripted pages before zoning boards and code enforcement hearings keeping a lot of solicitors and independent engineering consultants in business. An older version of the EIA code, (EIA RS-222-C) used to express *wind loading in pounds per square foot of pressure applied against the full height of the tower.* The fact is however, this pressure may vary at different height levels along the tower itself. It could be negligible for example at the base of the tower and up to 30 psf at the 300' foot level. Not only that, it may vary in different parts of the country. Subsequent revisions to the standard merely reset this reference height to 33 feet above the ground for a time. The standard was modified several times until the current EIA/TIA-222-E was released which supposedly represents the best technology to date. It is important for you not to entertain any "other" definition or standard than those currently accepted. The reason is that previous versions of the standards are virtually impossible to correlate with each other with any certainty which can be understood by the regulating authorities.

5.9 Engineering Diagrams and Blueprints

When ordering a tower from a catalog check the prices for engineering diagrams and prints. In some cases these can be provided *with* the order or they may be a *separate cost* item. Most manufacturers will not loose your tower order for the sake of picking up $1500 dollars worth of prints. But many times users will require the prints to see if approvals can be obtained prior to ordering the tower. The tower manufacturer can often provide similar installation diagrams (not your project) as a starting point, but when you get down to the actual construction diagrams, the order should be entered before your can expect to get the custom prints. Negotiate! It goes both ways. Most zoning boards do not need your actual prints if your tower is nothing out of the ordinary. In fact the least information and detail you present at this point will be to your advantage. Do not specify manufacturer nor show potential loading nor base details.

An example. You may wish to install the base for a 190' lattice (self supporting) tower. You may also wish only to seek approval for a 120' structure. By constructing a 120' tower on a 190' foot base, you may easily add seventy feet (70') of *steel* at a later date. You can actually order the tower designed for this extension. It is a lot easier to modify an existing structure than to replace it or apply for the entire structure height if controversial, at the onset. It's easier on the pocketbook too. Use your understanding of the zoning board *hoops* you'll have to go through as well. If your zoning board hasn't got around to setting up an antenna structure ordinance yet, you may want to slide through prior to enactment with all you can get. On the other hand, if your board has been hardened with previous presentations by the big boys, (often a ploy to make it difficult for additional users) they might be less objecting to a preliminary design with some expansion wording skillfully inserted into your application which *provides for future potential use while insuring that a need for additional towers will be minimized"*. If approved, you can return to the board in a few years for the extension. Things become a bit easier especially if the board is a new one and figures they didn't make the original decision.

5.10 Site Selection

In Chapter Two site identification was emphasized. In this section we are more concerned with the quality of the site. The first point is that there should be sufficient land allocated for the site and its future development. If the land is your own this may not be a problem. However if it is your intention to lease or option land, be sure to provide sufficient area for the type of tower you intend to construct, any restrictions such as fall zones which might be imposed, ground construction for shelters roads and right-of-ways. Other important considerations are soil type, land topography, the possibility of nearby natural or man made structures distorting your client's radiation pattern and location of the nearest airport.

Soil Type - Remember the "grassy knoll" on your property. Could that be an ideal location for the tower? Well maybe! It is a knoll because the geology under the top soil could be solid rock. There are mounting systems which mount into rock but the preparation of the base can be more expensive than the tower itself. Likewise, if the land is swampy, you may have an excellent ground system but no place to install a stable concrete tower base. If upon excavation you hit ground water or a water table before reaching the required base depth designed for installation, concrete will not cure properly and extraordinary installation conditions will need to be applied. Moving the tower location a mere hundred feet (100') could in many instances solve both of these problems and will have little effect upon the ultimate use.

Land Topography - Sloping hills, abutments and irregular terrain could create some unusual base considerations. It's not that a tower base can't be installed under these conditions but it makes equipment used at the site more difficult to operate and in some cases the land must be surveyed, leveled and cleared in order to provide the installation anyway. Here is a another true life example. While designing a microwave control station at a new police barracks, the tower height was calculated, the tower ordered and delivered to the site. Unknown to the engineering department was the fact that the contractor "leveled" the land for the building construction pushing about 20 feet of ground away from the tower location. When the tower was installed at the specified location the path for the microwave was no longer established, a mystifying development until it was learned what had happened. Fortunately the tower was expandable and the situation was corrected quickly. Even though the tower wound up being at the same height above average terrain (HAAT) it was now twenty (20) feet higher than that which was approved by the township. The matter was quickly resolved. Although situations like this don't happen everyday it is a clear example of what can happen if you do not keep track of the effects of everybody else's work at your site.

Nearby obstructions - Positioning your tower near a billboard, tall trees, another tower or even a building can create shadowing effects blocking portions of the coverage for your client. Although this is sometimes done on purpose, the trick is be out in the open as much as possible. Another favorite recommendation of zoning boards is to allow you to put your tower (usually on their land) in a tower cluster or "farm". Separation is required not only for pattern distortion but also for adjacent channel interference. When this type of interference is encountered it is better to separate vertically than horizontally. It is easier to manage interference on your own tower than to try to get someone on a tower within close proximity to correct a problem effecting you. If possible, stay away from antenna farms. They represent competition that you do not need (by providing clients with alternative sites at the same location) and present a host of other problems you just don't need to deal with at this point.

5.11 Tower Type

By now it is obvious that there are many different types of towers and antenna support structures which can be utilized. The decision to use any one type of tower is based upon many of the considerations already outlined. In many respects economics will be the underlying consideration. As a generalization, guyed towers represent the best tower for the money if the additional ground space is already available. Guyed towers can be erected at just about any height up to the super tower status. They can be easily loaded and expanded and require a bit more maintenance over their useful life. Lattice towers are self supporting and generally are practical in design to about three hundred feet. They are designed for a fixed loading and may not easily be expanded if not provided for in the original design. The monopole by far is the most expensive of the three and represents the least amount of multiple purpose application although monopole technology is improving with additional mounting options and loading designs being provided every day. Height is also a consideration for a monopole. The taller the pole the higher the ratio to overall cost especially at heights in excess of two hundred feet (200'). The problem with the monopole is that it is perceived as being the least *objectionable looking* structure when seeking approval. Accordingly it may be the only type permitted other than stealth systems.

And there are of course several variations and combinations of the above. Wooden poles have their application discussed in Chapter Three. Multiple close spaced guyed towers, or "doubles" were popular in the early 50's and provide greater loading if the *face* of the two towers is oriented towards the service area. Cross pieces between the two guyed towers positioned say twenty to thirty feet (20' to 30') apart provide a substantially greater mounting opportunity for a larger number of clients.

Conversions are also very attractive alternatives to commercial towers. An old fire tower, observation tower or similar system can be converted for antenna installation use. Usually a conversion took place prior to building code applications for tower structures. A local welder would merely *scap* iron and reinforced plates, here and there, erecting a base above ground after which tower members or sections would be attached. Some conversions are pretty creative but by today's standards would be difficult to obtain approval to modify let alone operate.

Tower *buildings* were also popular during the early 50's. It is basically a *tall* steel girder reinforced cinder block building with mounting opportunities and floors for equipment mounting. AT&T-tm used these buildings commonly for their mountaintop microwave systems for years and many are still in use today. From time to time and mostly due to satellite and fiber-optic technology a few of these sites have been declared surplus and sold to private investors. They are very nice installations even today but somewhat impractical under modern construction techniques and cost limitations.

The following discussions cover the typical tower systems you are likely to encounter:

Guyed Towers. A guyed tower requires enough ground space to position supporting *guy lines,* the number of which is dependant upon height and loading, typically attached to three (3) anchors, each of which forms an equilateral triangle. With the tower positioned in the exact center of the triangle, the *distance from the base* to the anchor point is approximately 70% of the tower height. If the tower is to be used for larger microwave dishes especially near the top, eighty percent 80% may be the rule. Each of the three (3) guy legs in this case would be approximately 125% of the tower height. This is a rule of thumb and can certainly change a bit but in general it is a good indicator of how much ground space will be required when designing a guyed tower system. As a practical example, suppose you intend to construct a two hundred foot (200') guyed tower. If the base is located at the exact center of an equilateral triangle, each of the three (3) sides of the triangle (guy lengths) would be two hundred-fifty feet (250'). The distance from the tower base to each of the anchor points would be one hundred-forty feet (140'). The *area* of the ground space needed (one side squared times 433) would have to be a minimum of twenty-seven thousand sixty-two square feet, 27,065.5sq.ft. or slightly greater than six tenths 6/10ths. of an acre. (27,065.5/43560 - where 43,560 is the number of square feet in an acre).

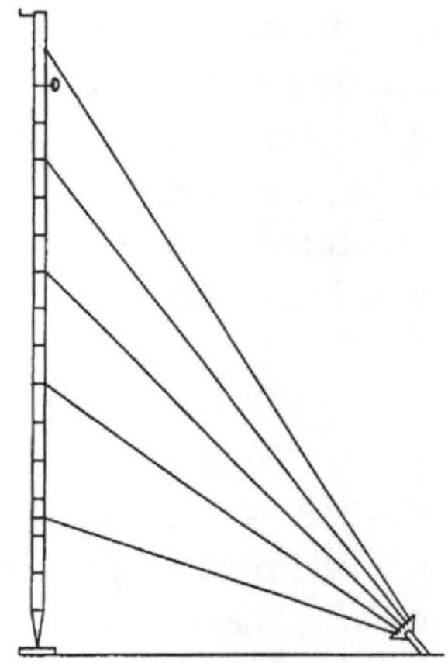

FIG. 5.11.1
GUYED TOWER

FIG. 5.11.2
ACTUAL GUYED TOWER

The above example and ground space plot may be altered to suit existing land conditions and specific anchor points so the land does not necessarily have to be symmetrical. If guy lines are moved closer to the base of the tower to use less land, greater forces will be applied to not only the resultant force "pulling down" on the tower (requiring a stronger tower) but to the guy lines themselves and their respective anchors. There is a point of diminishing returns where less land use results in greater construction costs to achieve the same loading. It is not uncommon for these percentages to be reduced to forty percent (40%) to achieve the same standards, but the trade off could be substantially greater material cost. Use a professional engineer if your design is any thing out of the ordinary.

Self Supporting Towers. If the design calls for a tower height in *excess* of one hundred feet (100') a guyed tower is substantially less expensive to construct. The availability and cost of the land is the limiting factor in this decision. Self supporting towers, however in heights of up to about three-hundred feet (300') can be erected in several configurations allowing substantial loading capability. Taller self supports are possible but may be cost prohibitive. Triangular and four point bases are provided by many tower manufacturers with several different types of construction. Tubular steel pipes, solid rod and angle-constructions techniques are employed in differing designs. Essentially the tower is delivered with a bucket (a rather large bucket at that) of nuts and bolts and skids of pipes or members of various sizes pre drilled and ready for assembly. All that is required is the diagram and a couple of heavy duty torque measuring socket wrenches. Once the base and the preliminary foundation sections have been installed, are plum and the concrete cured, construction may begin. Sections may be assembled on the ground and hoisted into position using a crane or gin pole. Or the tower can be assembled in the "air" one piece at a time.

FIG. 5.11.3 A 300' SELF-SUPPORTING TOWER

Angle (iron) stock towers pick up a bit more wind loading than solid rod tubing types but are less expensive. The problem with solid rod construction however is that it is substantially heaver than angular steel or hollow tube tower members. One advantage of solid construction is that corrosion *from the inside out* is eliminated. Because sections are pre welded, drilled and flanged, erection is a lot easier and the structure is easer to paint and maintain. Hollow round tubes require that *drain holes* be made in the base to allow water entering the inside to drain out. Should these holes become blocked with rust particles or dirt water can build up in the leg, freeze and literally burst the tubing. Angular steel is the least expensive to produce but is difficult to paint if required due to the increased irregular surface area. In addition since all the members are all bolted together, routine re-tightening is required. It also takes a bit longer to erect these systems. The wind drag on this type of tower is also greater than the tubular types. The decision to select any of these methods of construction is of course a trade off in the cost of the *steel*, erection and maintenance. In general however, solid tubing for the base and graduating to hollow tubing for the upper sections is about the best of both worlds and provides the greatest versatility and utility in multiple mounting applications.

As you observe many self supporting lattice towers and monopoles in service today the first thing you will notice is essentially only one user has been installed. Although this trend is changing, many of these towers were constructed with single applications in mind and generally can not be loaded much beyond what is observed. They were what I call *minimum bid single application towers* erected at a time when communication companies had no clue that there would be an additional four to seven companies in the same market. As a result, these towers either require *beefing up*, replaced or can not be used at all without jeopardizing the loading of the existing carrier's facilities. Re-engineering these facilities is under way but in some instances the engineering can cost more than a tower replacement. Should the tower be in a *grand-fathered* zoning region, the carrier may have no choice but to go with what is standing. In addition the height of these towers meets the needs of the existing carrier only, not a new carrier who might need a higher HAAT than that which the tower provides. There are a lot of factors effecting the reuse of existing towers but since the rents are the same, *your position of owning a tower with substantial capacity near a tower with no additional capacity can be a prudent move*. But you must be sure that the existing facility will not be upgraded in the future or your competitive edge will be neutralized.

FIG. 5.11.4
TYPICAL TOWER "ZONE"

Site Savvy

> Building your tower near a carrier's single purpose tower which can not easily be augmented could be a good strategy.

Monopoles - A relatively recent development, the monopole is a single element, flanged or sectioned concentric pipe, bolted to a huge concrete base and extending upward to a desired height. Transmission lines are pulled up *inside the pole* and emerge at the top and bottom through little pre-cut doors for that purpose. Because the pole section diameter decreases in circumference as the pole tapers higher and higher, the maximum number of transmission lines stuffed inside the monopole is entirely dependant upon the diameter of the pole at the top point where they

Fig. 5.11.5 Sabre Monopole with three platforms

emerge. Usually a pedestal or platform is erected at the top of the pole. On some models this platform actually extends above the top of the monopole. Antennas are attached to mounting pipes on the pedestal railings. The platforms are triangular and can be shifted around on the pole to establish any required face directional orientation. A small *catwalk* enables the technician to walk around to each antenna for testing or replacement. The climb to the top is accommodated by foot holds attached or welded up the side of the pole. A safety cable also runs up along the same side.

5.12 Types of Antennas, Patterns and Directivity

A monopole is relatively expensive when compared to other types of support structures. Also, loading considerations assume that most of the antennas installed on a monopole will be small high frequency "sticks", panels or sectorized antennas. Because these antennas have a very narrow beam width which directs RF energy forward, the rear of the antenna ("off the back") will not be distorted by the pole or mounting member itself. By clustering several 60° degree beam width sector antennas around the platform, there is little or no pattern distortion because these antennas are *unidirectional* rather than *omnidirectional* which would have to be mounted on the top of the monopole out in the clear. If microwave dishes are also added to the monopole at a high pont, the stability of the pole must be greatly increased to prevent what is termed microwave *sway and twist.* This is the tendency for the tower to actually rock slightly under severe wind conditions. An inch or so sway usually won't effect the performance of a panel or omnidirectional antenna, but it could significantly alter a microwave dish path causing reduced signal strength or loss of signal (LOS) all together. It is obvious from this situation that microwave paths established using monopoles work a lot better if the dish is not located near the top of an *underrated* pole. Positioning the microwave dish further down the pole might prove usable due to lessening effects of *sway.*

There are five (5) basic types of antennas which mount on towers or supporting structures.

The Omnidirectional Antenna - This antenna consists of either a single quarter wave element or several quarter wave elements phased together and mounted inside a tubular "*stick*". At VHF and UHF commercial frequencies these antennas may vary anywhere between about two to six inches (2" or 6") in diameter and from four to twenty feet (4' to 20') in length. Substantially longer "sticks" if not mounted at the very top of the tower, usually require two (2) mounting points. One at the base or feed point of the *stick* and the other at the top of the *stick* to prevent it from swaying in the wind. The top support is of course non metallic as a metal support pipe at this point might distort the antenna pattern. Stand alone *sticks* are substantially heavier to prevent swing when no top support pipe is used. Thus there may be a weight difference in the two types of antennas depending upon where they are mounted and their overall length. Short *sticks,* four to eight feet in length (4' to 8') don't swing much in the wind. Also because their vertical beam width is wider, even if they did swing several degrees, the pattern wouldn't be effected greatly. Twenty-foot (20') *sticks* however, swing quite a bit unless the support tube is a lot heavier or the *stick* is braced at the top. Usually longer antennas have substantially narrower beam widths, sometimes as narrow as 7°, a swing of a few inches can lift the radiation pattern right out of a coverage area. Keep this in mind as it also effects loading and pricing.

A paging antenna for example may consist of several co-linears *inside* one (1) single **tube** all fed by one (1) transmission line. In the PCS and cellular service however each antenna is usually fed separately requiring a single transmission line for each radiating element. *Transmission lines substantially add to tower loading* by adding wind drag, additional surfaces upon which ice can form and if improperly bundled or run up the sides of the tower can even restrict additional antenna mounting. This is due to mounting positions or lines getting in the way. On FAA regulated towers, transmission lines might also need to be painted *red & white* every twenty feet (20') or so. Because transmission lines run inside the monopole they would not have to be painted using this support system.

In some cases, radiating elements known as "hairpins" are not contained within the *stick* but strapped to the outside of a tube and fed with a coaxial harness or *power divider*. This type of antenna is very versatile because each "hairpin" or dipole can be oriented to shape the radiation pattern in an infinite number of ways without being at the top of the supporting structure. Because the supporting pipe is at ground potential (hopefully), the radiating elements are not as vulnerable to lighting as they would be using the co-linear design. These antennas are practical to about 900 MHz. and are usually fed with one single transmission line. If these antennas can be mounted far enough away from the antenna supporting structure on extended brackets, the blocking effect of the tower is reduced significantly, *but loading is increased.*

Your client will sometimes object somewhat to being positioned anywhere but at the top of the tower and will usually insist upon not being *too close* to anyone else's antenna. Pattern distortion and RF interference is the concern. What you need to know here is that pattern distortion and RF interference is a function of frequency and may or may not occur as predicted by your client's adherence to a rule of thumb. In addition the effects are predictable, measurable and correctable. Which means that if concerns are voiced you must contractually control your options to place clients at their optimum location consistent with *your* objectives for co-locating additional customers. This concept was previously explored when the effects of permitting a communication carrier the option of mounting sector antennas on a building elevator penthouse wall to the exclusion of any additional client was explained. Failing to take this into account will result in significant capacity reduction on your tower limiting your revenue potential.

Another point to consider is that if one tower face or building wall points towards a major population density (service objective) which is the target of your client's radiation pattern, everyone will want to mount on this face to the exclusion of the others causing undue loading bias. A predictable tendency not easy to correct after the fact. Thus positioning of the tower legs and mounting brackets in the initial design can substantially increase the loading in a desired direction if it is considered prior to installation. This effect can creep up on you without notice so care must be taken to visually inspect a series of installations.

Sector Antennas - Sector antennas solve many pattern design requirements for a host of telecommunication services. Consisting of moderate to high gain antennas contained in radomes, both the vertical and horizontal beam widths can be controlled or adjusted. It takes a series of three, six or nine antennas (literally wrapped around your supporting structure) to provide the desired coverage in vectors of approximately 40° to 60° degrees. In addition, sector antennas can be tilted down or up in the vertical plane to shape a pattern with a great deal of precision. Add to the fact that each antenna radiating element is fed with its *own separate* transmission line the selection of which sector to use at what time, and the power applied to that sector becomes a controllable parameter in the operation of the system. Sometimes referred to as *intelligent antennae*, better service, including service to more subscribers, better range, less noise, fewer call drops, and overall improved service will result. In addition, because these systems are wide band in nature *more than one service can utilize the same array at the same time*. A point which requires contractual consideration if you do not know what is happening.

When considering the effect of such an array upon your supporting structure it is important to note that these antennas do not need to be mounted in the clear nor at the top of your tower (if the HAAT at the mounting point is sufficient) as they have virtually no radiation or reception off the back or mounting pipe and can literally be fastened directly to the tower leg itself without pattern distortion. Secondly, remember that each antenna must be feed with its own coax cable and because the frequency is in the 800 to 2000 MHz. range, these cables are going to be quite large in diameter (typically 1 5/8" or greater). Finally, it is not uncommon for sectorized systems to evolve, so be careful to provide for this expansion at the onset. If a telecommunication company starts out with several *sticks*, but indicates that it intends to move to a sectorized antenna(s) at a later date, they may try to lock in expansion rates based upon the initial configuration, not the supporting hardware and structure involved by maxing you out at nine antennas and nine transmission lines.

Microwave and Satellite Dishes - The third type of antenna finding its way on you're supporting structure will be microwave dishes. These parabolic antennas can be very heavy, very large and when mounted inside fiberglass radomes (protection envelopes) create a tremendous amount of wind loading. So much so that many microwave dishes have built in heating elements to melt ice build up. Ice will distort the pattern and lessen path reliability so the heating elements are not primarily added to decrease your loading but to maintain dish reliability. This also means another conduit suppling AC power from mains running up your tower to the dish which is often overlooked when pricing the customer's needs and calculating tower loading. In addition, it is a common practice for microwave users to want to install an *ice shield* just above the dish which is between four to twelve feet (4' to 12') in length in order to protect the microwave antenna from ice slabs falling down from above. An ice slab may not hurt the antenna much but it sure can jerk a dish out of alignment interrupting its path. These shields also add substantially to the loading mix and must be taken into account in not only the design but the pricing of the installation. Microwave dishes should be reviewed by a professional engineer for a statement of their loading effects prior to installation and pricing.

Another common dish used by clients are single channel or sub-carrier channel VSAT systems which distribute paging traffic on a receive only basis to the transmitter located at your site. Large paging companies up-load paging traffic to a satellite, then distribute a single channel loaded with data to multiple transmitter sites over large coverage areas. The advantages are that telephone lines are not required and that all the transmitters in the network will receive their data simultaneously. The disadvantages are obvious. A micro meteor can wipe out the satellite and put 40 million paging subscribers "off the air" without a backup distribution plan. However the use of a satellite distribution system allows many paging companies to simulcast their paging traffic on all their transmitters at once.

Should multiple transmitters with overlapping coverages be on the air at the same time, a great deal of interference would result if the data and the frequency is not aligned carefully. In fact if *simulcasting* is done properly two overlapping signals might actually enhance each other, improving the overall paging coverage. Most of these dishes are small aperture, about three feet (3') in diameter, and are fixed (aimed) at geostationary satellites in orbit. If your site does not have a clear path towards the Western sky, low to the horizon (especially in the East) it may be necessary to mount these dishes on a tower leg.

When designing your antenna site be sure to consider aligning your shelter or building to permit mounting several of these dishes along the roof rather than using your tower. They need only be higher than the highest predicted snow fall and do not always have to be mounted on your tower.

GPS Antenna - Another small antenna utilized by communication clients is the GPS antenna. The operation of the GPS as a navigational and locating device was reviewed in Chapter Three. However another function of value to the communication carrier is the reception of an accurate time base. National Institute of Standards and Testing (NITS) timing signals are used to calibrate transmission frequencies and other system timing devices and electronic clocks crucial to proper data transmission, logging and frequency accuracy. These antennas are very small and do not need to be mounted on the tower. They are often overlooked in the facilities description in the site agreement.

Ground plane - Some antennas require sufficient space for a ground radial system mounted just below the radiating element. At lower frequencies these radial lengths can be significant and cause pattern distortion to other types of antennas mounted nearby on your tower.

Ground plane antennas are usually unity gain, are omnidirectional and can be driven at high power levels. They are used primarily at Low Band 35 and 45 MHz. paging applications, rural and some UHF applications. Other variations of this antenna beam the rf energy directly up in a *cone pattern* used in some aviation applications. In general this antenna is not as common as the other arrays and will more than likely not be a significant factor to your overall plan.

Yagi and Other Specialized Antennas - Many local communication clients may use in-band, 75 MHz., 450MHz. or 950Mhz. remote stations to control their base stations or repeaters located on your tower. These antennas are highly directive and point specifically to the remote station. The only requirement is an unobstructed path to a point (in a valley below for example) where the reciprocal low power control station is located. They look a lot like small TV antennas consisting of a boom with multiple elements attached along the same plane. There are several types of Yagi antennas that create some concern for tower loading. A three to four element yagi for 75 MHz. applications, any fixed receive antenna used in the FM Broadcast Service (FCC Part 73) where a translator might be employed and finally any CATV application where receivers for TV channels two through six (2 to 6) are used. These Yagi antennas can be quite large, heavy duty and must be mounted on tower legs at various points. When icing conditions are encountered, these units can also get quite heavy and must be considered in your loading considerations. They are essentially receive only systems (except for some 72-75 MHz. applications) and use much smaller transmission cable because the frequencies utilized are lower thus reducing the loss considerations encountered at higher frequencies.

FIG. 5.12.2 YAGI ANTENNAS ADD SUBSTANTIAL LOADING TO THE TOWER

In addition to Yagi antennas are special application antennas such as FM broadcasting or VHF/UHF TV transmitting antennas. In most cases these antennas require a top un-obstructed mounting, however there are many designs which work well when side-mounted. Usually a tower will be constructed for this application in the first place so it is unlikely you will have a commercial broadcaster as an add-on client. However, with the introduction of translators and low power TV, MDS and FM systems, add-on commercial broadcasters may still show up at some point in time. With the recent release in HDTV applications, substantial power will be required to a much heavier TV antenna. Thus a whole new class of tower is being considered. A *cooperative super tall* TV tower may be in the offering which might eventually free up an existing TV tower which incidently needs to be used for a time in supplying conventional TV service.

A word about high power TV and FM antenna systems. Should these systems be mounted at the top of your tower, some consideration should be given to the non-ionizing radiation hazard presented by these strong RF fields. Essentially having the same effect as a microwave oven on human tissue, the skin will begin to heat up slowly raising the body temperature. If the signal levels are high enough and the exposure is long enough the body temperature could be raised to a point where unconsciousness could occur. Not a good thing to happen to a technician who is climbing your tower. Notification on the tower at the point where excessive RFR is encountered is not only responsible it is the regulation. We will have much more to say on this subject in Chapter Nine.

5.13 Significance of Positioning and Combiner Technology

The type of antenna, mounting hardware and how it is fed to the transmitter becomes a significant consideration when determining tower loading vs. pricing. A single "stick" fed with one transmission line, may in fact be a combination of several antennas within a fiberglass tube or mounted on a pipe. The antenna may be used for both transmission and reception simultaneously with the use of special duplexing equipment or a two-conductor transmission line in what appears to be a single cable.

Some system operators, realizing that reception systems can be contaminated by transmission facilities operating too close reserve the upper most portions of their tower for reception systems allocating a lower portion for transmission and microwave applications. This would only be possible if a good deal of the tower height is at usable HAAT and not just there to position an antenna above the tree-line. For example, if a four hundred foot (400') guyed tower on a hill top provided clear coverage above the trees from the one hundred foot (100') level to the horizon, then about three hundred feet (300') would be usable for various antenna mounting applications. Since line-of-sight from the 100' foot to 400' level would be almost equal, it wouldn't matter much where antennas were mounted. In this case the site operator designated the top one hundred feet (100') on the tower for receive only applications and transmitting antennas from the three hundred foot (300') level down to the tree-line.

As can be seen from this example, a closely guarded secret of tower height does not necessarily have to be to divulged to achieve a single desirable HAAT for one application. It can also provide a usable *range* for many co-located applications. A concept not fully appreciated by zoning boards in the past.

Should the height of the tower be fixed to a HAAT for a given application, some services might still work fairly well *beneath* the design height. If space is at a real premium, a site operator may install a single wide band antenna at the top of the tower and feed it with a high quality low loss single transmission line into a *combiner*. **www.wacomprod.com** Subscribers desiring a receive application (in addition to their transmission facilities) merely "plug" their systems into the *combiner* and share a common antenna.

Combiners also can be applied to transmission facilities as well. So theoretically for a group of 900 MHz. paging clients, a single antenna could be used by everyone. There are trade offs. Power to combined systems is wasted in the process, certain frequencies are incompatible with each other and if any part of the system is corrupted or fails, all the users go down. Objections are very vocal in these systems and must be expertly managed one hundred percent (100%) of the time with no exceptions. There is just too much at stake. Where it works, for example, is on the runway of a large airport; a single tall structure in the middle of a metropolitan area, or in an industrial park that allows *only one* antenna! That one antenna, referred to as a *must have site*, can do a substantial amount of work under carefully controlled conditions. An RF consulting engineer can put such a system together and there are several companies that manufacture highly reliable *combiners*. **www.telewaveinc.com** The technology exists but it is not the preferred method of site utilization. However sometimes it's the only ball game in town.

5.14 Tower Lighting Systems

We will have a whole lot more to say about tower lighting systems in a later chapter but for loading considerations it is safe to say that the top position on the tower, conduits and heavy mounting accessories are all part of the FAA requirement when tower lighting is necessary. These systems can be extensive and costly requiring maintenance and monitoring with severe penalties levied upon the site owner if anything goes wrong. Towers are lighted because it has been determined that they represent a possible hazard to aviation operations. Fortunately the criteria under FCC Part 17 is fairly straight forward and there is not a whole lot of confusion as to what is required and how the system should to operate. In recent actions however, zoning boards (am I picking on them again?) in their zeal to "protect the public" have created some rather absurd regulations. In one instance there was that requirement to paint a monopole *sky blue* so that it would "blend into the sky". In another we were required to paint the PCS panels on just the antenna, *international orange* (apparently not aviation orange) to make the antenna "visible to a local ultra light aircraft operator" who used his back yard as an airstrip. Both *orders* were totally out of line with regard to FAA regulations but nevertheless were conditions in order to receive building approval. If by chance an aircraft or helicopter ever collided with the "sky blue" tower, I wonder where the liability would be directed? Or do zoning board members have *sovereign immunity*? At any rate it is best to comply with FAA regulations or at least obtain an FAA determination as to what regulations apply to your structure or building. They are obligated to provide a formal response to your formal application using their form whenever there is a question or an "order" relative to aviation hazzards related to your tower.

5.15 Intermodulation, Distortion and Interference.

Although as an antenna site owner/operator you may never be required to correct an interference problem arising between your tenants, you may be required to coordinate or negotiate a solution. Accordingly your site agreement should contain wording to the effect that you have this responsibility and right. However, usually it is the custom for any new entrant to your site to coordinate and pay for a solution to any problem which arises. The logic here is simple. If the new tenant had not installed their equipment at the site there would be no problem. Or stating it differently, if the new tenant were expelled, the problem would go away with the tenant.

In reality interference can take many forms. An example. If two tenants (number #2 and #3) each operate a paging transmitter there might be no problem at all until both transmitters are "keyed up" (turned on) at the same time. An *odd order frequency product* can be created which falls across the input band pass of another tenant; we'll call this tenant number #1. Because #1 has been there from the beginning, this tenant's control receiver is now completely useless. Who is to blame? Tenant #3 because he was the last to arrive at the site? Tenant #2? Now things get difficult.

Before you notify tenant #3 that he will have to leave your site or that his transmitter is to blame, you must make several determinations. First and foremost are all three tenants operating their systems in good engineering design? Is everyone using proper shielding, coax cables and grounding? Is the equipment modern and type-accepted? I operated a paging site a few years back which interfered with a nearby resident's TV set. No amount of engineering "tricks of the trade" nor devices solved the problem. The TV was an old black & white Silvertone-tm produced in the late fifties. The solution was to purchase the resident a new color TV which complied with current FCC regulations. The problem was never corrected but a solution was possible.

The next step is to understand what is causing the problem. Interference can take several classic forms which are well understood and pack a host of solutions to try. If transmitters are not properly aligned or are running excessive power they will spill spurious radiation out of the proper band pass causing interference. A device which allows a technician to actually *see* what is going on is called a *Spectrum Analyzer*. It will find the problem quickly, and put values on each parameter so that calculations and remedies can be tried. Are the antennas too close to each other? If the antenna is removed from tenant #1's equipment does the interference go away?

One of the most common causes of interference is intermodulation distortion (IM). When signals combine in a non-linear device they may create other undesired signals (small transmissions on other frequencies) which pass through the original transmitter amplifier tuned circuits and then are radiated out to an antenna. Often the insertion of a filter or other active device in any or all of the systems can attenuate one of the signals or both of them to the point that the offending signal is virtually eliminated and thus the problem is corrected. But non-linear devices don't have to be transmitters. They can be rusty bolts, dissimilar metals joined together, even a pile of old antennas and pipes lying out back behind the transmitter shelter. Fortunately the non-linear device doing the signal mixing, can also be located easily by using calibrated DF (direction finding) equipment or a tunable field strength meter. Can you believe that at one time I spent a couple of hours looking for a *non-linear translator* that turned out to be an old corroded door hinge? A shot of oil and the interference went away!

Another type of interference is caused by harmonic energy being improperly radiated by a transmitter which is incorrectly tuned. Usually this interference is a *direct multiple* of the actual transmit frequency or one of the intermediate frequencies used in deriving the output frequency of a given transmitter. This type of interference is relatively easy to deduce and correct as all the numbers must divide down to known frequencies.

Adjacent channel interference occurs when transmit frequencies of relatively high power are just too close (say within 100 KHz. or less) to a receiver frequency. So much energy is allowed to enter the receiver on a frequency which is close but not exactly on channel, that the desired receive frequency is *covered up* by the adjacent channel energy. Once again filters help quite a bit but nothing short of changing the frequency will completely solve this problem.

Co-channel interference and cross-talk occurs when two transmitters on or near the same frequency overlap into one receiver location. Either the transmission of one of the signals is too strong, or an antenna requires reorientation. Fortunately this type of interference is intra system and won't be a problem to other tenants. The problem however is that it may initially be diagnosed as another type of interference which gets everyone else all upset until the problem is identified.

Finally there is signal blockage which is caused when tree growth, or new construction or other new obstruction blocks a previously clear path creating a coverage problem. Most tenants will want exit clauses inserted into their agreements to allow them to move their facilities should this problem arise and not be resolvable.

You certainly do not have to be an RF engineer to manage these difficulties. There are many good texts which deal with the practical solutions to these and a host of other common and exotic interference problems. You must however keep the personalities involved calm and focused to act quickly to get to the bottom of the situation and make a decision. Failure to do so will result in your not only losing tenants but you will gain the reputation for a *dirty* site. That is the kiss of death to your marketing program and there is no need for it to happen. The site acceptance survey is one good way to screen tenants to be sure equipment will be compatible. It also provides data which can be used by your new tenant to compute probable and thus avoidable problems prior to installation. Once the problem is solved it merely goes away and everyone is happy again.

5.16 Tower Safety Equipment and Hardware.

Tucked in the back of almost every tower supplier catalog are products used for the safety and convenience of tower maintenance personnel. Towers usually have optional ladders and climbing tools as a part of the overall project which can be ordered and installed at the time of the initial construction. Most tower manufacturers have a good selection of safety materials and information concerning their products along with tips on tower safety for both installation and operation. Collecting these brochures is a good idea and lots of practical information can be obtained from them. Because you rent tower space you will more than likely have little control over who climbs your tower. You can however have posted tower climbing rules to protect you from apparent negligence. Hard hats for example *are an absolute must* when working around a tower because it is very easy for a technician to drop a galvanized *U-bolt* from the two hundred foot (200') level which can penetrate the human skull to about a half an inch or kill someone instantly. Rigging and safety belts should be sized for the man. Canvas buckets for tools and small parts, safety lines, slide clasps and pulleys allow ground crews to pull up and position heavy antennas during installation. If you indicate in your contract that experienced antenna crews must be utilized you at least will have set some standards for personnel climbing your facility. What you don't want to happen is an accident caused by the negligence or improper equipment on the part of a technician resulting in a liability lawsuit directed to you for negligence. A good defense against this occurrence will always be safe modern equipment and posted (and acknowledged) rules and regulations. And don't forget the all important waiver of liability clause!

FIG. 5.16 ORDER SAFETY BROCHURE FROM ROHN® AT 309-697-4400

CHAPTER SIX

6.1 Marketing Your Site

You have selected and researched your property and have determined that it is a prime location for an antenna site. You have communication company interest (maybe even a contract contingent upon construction by a specified date) for your roof-top antenna site or your proposed tower site complex. Now you are ready to open for business. You will also need an acceptable site rental agreement to sign-up your clients. Don't worry if you don't have one right away, communication companies will place one in front of you at the onset of negotiations. And by the way, don't be too quick to sign it until you understand what is happening. We will explore the site agreements, license and rental contracts in detail in the next chapter.

Unless you are aware of site acquisition activity going on in your area and have already lined up your first tenant prior to construction (the safest course of action) let us assume you are starting from *square one*.

6.2 How Site Acquisition Really Works

One of the easiest ways to market your site is to *turn it over to a site representation/management company.* There are many site management firms from which to choose. The advantages and disadvantages of using this technique are numerous. But for now let us assume you plan to represent your own interests. An understanding of site acquisition activities, the way sites are considered and selected can be helpful to your business plan. The basic principles and tasks of the Site Acquisition Specialist are the same whether you represent yourself or have a management firm perform the functions on your behalf.

In the early days of cellular, towers and buildings were selected in order to provide reliable coverage for *mobile* units. As portables replaced mobile units in popularity more sites were needed because the range of the portable is typically less than that of a mobile installation. Once again, reliable coverage was the primary objective in site selection.

As the number of subscribers increased, additional sites were required not for reliable coverage any more but for the *process of cell-splitting*. This is the technique of providing smaller and smaller coverage areas in order to serve more subscribers on the same frequencies.

Accordingly, reliable coverage, number of users and revenue potential were added to the mix in site selection.

This concept is explained by Gregory E. Sweet, *Acquire Telecom Services*, in an article appearing in *Radio Communications Report* (RCR). Sweet correctly observed this trend in companies that have been around for a while. He points out that qualified site acquisition people, acting as subcontractors for telecommunication ventures, *often know* before a site is even constructed that it will not meet all the network/marketing objectives desired. The reason given is that radio engineers who have been charged with the task of designing the network do so with great zeal but usually on paper. Radio engineers may have been given rather rigid criteria to work from and can be reasonably sure that what they design will if constructed perform properly from an engineering perspective. They also perform knowing that if what they design does not work according to plan, they might loose their contract, job or even worse, their engineering credibility. Difficulties arise when what has been designed on paper can not be completely constructed in the real world.

If a given area is being *engineered* the product is going to be a hypothetical set of overlapping coverage patterns on USGS maps. According to Sweet, there is a potential for some confusion here. Many of these maps are out of date and might show open land where newly constructed (or zoned) residential developments now exist. Each pattern will have a focus point, i.e. a theoretical point where the tower is projected to stand, marked on the map. Encircling this point is a stated deviation usually about a half a mile in all directions. The area scribed by this deviation from the focus point is referred to as the *search ring*. The engineer then turns these *search rings* over to the Site Acquisition Specialist.

The Site Acquisition Specialist, usually working on commission for completed site leases, trudges out to the physical location and finds a nice residential development full of NIMBYs slang for folks who yell *Not In My Back Yard.* Once I found a swamp filled with nasty snakes and poison ivy. The land owner is then located, a *pitch* (a.k.a. *chin boogie*) is made and hopefully a *contingent upon approvals deal* is cut. The exact location is returned to the engineer who reworks his *engineering,* determining the height of the tower required, types of antennae needed and its relationship with the adjacent *search rings* which may or may not have been acquired up to this point.

As can be imagined this process is cumbersome and filled with many difficult to control variables. There may be a conflict of interest or two, lots of paperwork and rework. All of this activity might not result in a completed tower or network being constructed on time or even be within budget at all. The radio engineer seemly works in the ideal realm while the Site Acquisition Specialist is left with the practical world. Often it takes a management override just to resolve the two points of view which too often are not resolved at all, but redirected to other site possibilities where the process starts all over again.

Taking the process further. Once the site is put under contract and approved by engineering the construction phase might take months, years or not happen at all. I was given a project to construct. The building permit was issued. All the steel and equipment were scheduled for delivery to the site. Everything was on schedule and well within the required *civil start* date on the permit. On the day construction was to begin, a small red sign was nailed on a utility pole at the site. It said "STOP WORK". The political meanderings behind this action notwithstanding, the project was dead! Now imagine what this does to the engineering network design, the time table, legal costs and the project costs for construction; the Site Acquisition Specialist's commission not to mention the immediate problem of where to store a 150' monopole for a while.

FIG. 6.2 AUTHOR FOUND A "STOP WORK ORDER" ON TOWER INSTALLATION DAY

Suppose you are a multi-site owner/operator. Suppose your hard work and investment has provided the "only corridor of reliable coverage" along a fifty mile (50) stretch of an Interstate highway. You have six (6) towers in all, spaced at about eight mile (8) intervals. Each of your sites, in addition to serving the highway exhibit a slight overlapping coverage. Every other tower *sees* each adjacent tower in order to accommodate any desired microwave application. Each site has plenty of antenna loading capacity, 400-Amp service, telephone lines, emergency power, a paved road, security and air conditioning. Your product offering doesn't get much better than this if you are in the site business. After a time you are told by a communication company consultant that. . . "*sorry, your sites do not fall into our search rings*". What is the problem?

The first consideration is that a lot of consultants get compensated by *the project*. If your network/system is ready to go, that might be six commissions not received, six pro forma engineering studies not required, six towers not constructed. A lot of savings for the principals & stockholders, but they might never know about your system. Their consultants might *steer* them intentionally or innocently to a different configuration for any number of seemingly valid reasons. Yet if only one (1) of your existing network tower sites was selected, the other five sites would fall into place and they could also be utilized for the same corridor objective.

What would keep a communication company from selecting your existing network? Well, first and foremost, good engineering design. Secondly, your rental price and their urgency of need. If your rent is too high or they don't need the coverage right away, they may attempt to construct around your system or take an option until they develop the cash flow to support the expenditure.

Many new PCS carriers do not have the resources nor the time to build-out in the same manner as their cellular counterparts have done over the past fifteen (15) years. However management consortiums and joint ventures constructing PCS systems have been tempted to sub-contract everything from network design to system construction, operation and marketing for their investors. Since each discipline tends to have a somewhat narrow view of its own functional responsibility, there is bound to be inter-functional rivalry if all the sub functions have not been assigned to the same company. Unless there is competent project management, these rivalries can be intense, with blame for missed target dates and cost overages running rampant throughout the project.

Ideally, Sweet suggests, that site acquisition and zoning efforts be initially directed towards *existing facilities*, roof-tops, towers and friendly zoning boards *prior* to engineering system network design. He also suggests that existing sites which can be utilized immediately, be presented to the engineering consultant with an order *to make it work*. Full consideration must be given to the revenue generating potential of the site location rather than the technical coverages the site might need to fulfill as a part of an idealized or hypothetical network. Radio engineers may not have this capability. This is a job best left for econometrics or marketing expertise.

Finally, a site doesn't necessarily have to meet a 100% network objective. Sweet suggests that sixty to ninety percent (60% to 90%) might best serve company needs. Accordingly the radio engineer needs to have this information early on.

What happens when the company simply can not obtain a few of the sites in the hypothetical network plan of overlapping coverages? Initially, reliable coverage can be provided for a time prior to subscriber build-out with what is called a *boomer-site*. Here is an example. A mountain top tower is fitted with a PCS base station which serves the build-out staff in the market prior to *turn-on*. Commercial service using this site would be impractical because its coverage is just too large and it can only serve about 24 users at one time. However, if the network calls for three or four fill-in sites, low volume is anticipated, and only one can be obtained initially, a nearby *boomer site* can be operated at higher power for a time to provide the reliable coverage sought but without the capacity to serve more users. Fortunately it takes time to build up a customer base to the point where a *boomer site* either has to power down and become a smaller part of the network or is abandoned all together. But gaining time as well as cash flow, is a major PCS objective. In this way the site acquisition and construction process can run simultaneously with initiation of local service.

Sweet concludes with recommendations that communication companies *work together* on common sites, build extra capacity into existing sites under construction in anticipation of co-location and promote a *cooperative spirit of public relations* and education in order to make zoning and local approvals much easier. His article was written in September of 1996, more than enough time to measure the impact of this recommendation and it's effects upon the industry.

There is no question that cooperation and co-location is not widely taking place. However, generally it appears to be a bit more cautious than that envisioned. One reason is that any cooperation or co-location is going to take place between fierce competitors perhaps to the exclusion of other competitors in the same market. For years some cellular companies refused to permit the co-location of any other communication company facility on its tower infrastructure. Some PCS's were annoyed by the practice hardly accepting the reasons given by the cellular operators that their facilities where not designed for more than one user or they didn't have any additional room for PCS systems. When *ordered* to provide co-location by local zoning boards, some cellular companies acquiesced by issuing open acceptance of co-location policies while introducing higher than market (and seemingly prohibitive) rate structures. During this time, PCS operators literally had to start from scratch. They could not depend upon successful negotiations with cellular companies with any certainty which seemingly took months to conclude one way or the other. Because a cellular company might be totally capable of serving an entire market with its infrastructure, if it did allow co-location, a competing PCS could be on the air with an identical coverage foot print in a matter of months. For the sake of argument, suppose a cellular company operated a market with 140 towers. It cuts a deal with a new PCS to provide tower space for $2000 per month per tower. That's $3.36 Million a year in revenue for the cellular company. It is also a savings of about $30 Million for the PCS in not having to construct the same facilities. The PCS launches its service and takes 20% of the cellular company's revenue in *churn*, i.e. the other guys customers, amounting to about $14 Million. The net loss to the cellular company is $10.6 Million per year. If the deal hadn't gone down, it might take the PCS three years to accomplish the same objectives resulting in the cellular company's collecting $31.9 Million revenue it wouldn't have. But now don't forget the capital outlay and debt reduction now imposed upon the PCS to pay for the equivalent number of towers it was forced into building without the cooperation.

To keep up appearances however, from time to time a cellular build-out objective and a PCS tower location wind up being in the same search ring. It is then that a specific deal is cut between the parties. However the overall effect upon either company's marketing objective may be negligible, but both companies can claim *co-location cooperation*. Agreements to do this sort of thing are in place and attempt to serve as a good faith response to zoning board requirements that *all existing avenues to obtain sites must be exhausted prior to consideration by the zoning board for approval of any new facilities*. Tacit cooperation at best!

Now that you understand some of the many forms and intricacies that this historical process has taken you will be in a better position to meet the objectives of all the parties concerned. Serving as an arbitrator it may be to your advantage to bring the parties together, in neutral territory, (ideally yours) providing what they need but what they might not be able to initiate on their own.

But you could have other competitors in this strategy. One of them may very well be the local municipality itself. In the name of government, the concepts of competition may be compromised by the position that the zoning board can exert over the situation through its zoning and code enforcement powers. The risk is great that such a process may have elements of restraint of trade, undue government manipulation, unfair practices and interference in free enterprise. These issues have yet to be challenged with any great impact but will perhaps surface when PCS companies, who in good faith have bid outrageous amounts of money for the privilege of their licenses, are unable to complete their systems due to unreasonable local government interventions and restrictions. Keeping current in this controversy may provide you with opportunities to achieve very profitable objectives.

6.3 The Advertising Program

The first thing to do is to get your site listed in the two major site resource catalogs. They are:

1. National Communications
   Site Directory
   Intrafamily Communications, Inc.
   PO. Box 6093
   Freehold, NJ 07728
   908 462-5964 Voice
   908 308-4633 Fax
   http://www.intrafamily.com

2. Fryers Site Guide
   & Electronic Data Service
   84 N. Lansdowne Ave. #300
   Lansdowne, PA 19050
   610 284-9289 Voice
   610 622-1697 Fax
   http://www.fryersguide.com

Sites are identified in specialized listings in a standard format. Typically a site listing for a building would look like this:

> **NV, Las Vegas** AGL: 400' AMSL: 2910'
> 036:08:24.0 N 115:10:05.6 W Tel: 702 220-5555
> **Hollar Communications** Norwest Casino & Hotel in Las Vegas. Close proximity to major tourist areas and convention facilities with excellent in-building penetration. Tight security, 24-hour access, stand-by electrical, air conditioning, phone lines and parking. Clark County.

For a tower location:

> **PA, Harrisburg** AGL:190' AMSL:1170'
> 40:15:52.93 N 76:51:53.82 W Tel: 717 555-1212
> Blue Mountain overlooking State Capitol Harrisburg PA.
> Self supported 190' tower, block building, paved road,
> good security, emergency power.
> Dauphin County. Hollar Communications.

There is a presentation point which needs clarification. Published tower or structure height designations follow an *unusual format*. Normal practice is to give the structure height above ground level as the AGL in feet. However, when expressing the overall height, the ground elevation level above mean sea level, AMSL is *added* to the above ground elevation of the structure to obtain the *published* AMSL in feet. In the above example for the self supporting tower, the *ground* elevation above mean sea level would be 980' AMSL. However, the practice is to *add* the structure height above ground to the ground elevation above sea level producing a new value for the AMSL as 1170' (190'AGL plus 980'AMSL).

Keep in mind that this new value expressed in various publications is not too hard to visualize when considering a roof-top because most mounting heights are all going to be about the same level, i.e. spread out over the roof area. However on a tower, an antenna can be mounted usually anywhere from the top. . .down in which case this interpretation of the AMSL would vary. In addition, for an existing tower in use, it is quite likely that the top sections are already rented and the AMSL given might not be available. The published value for the AMSL might now be misleading to anyone other than consultants who are familiar with the protocol. Since you will be preparing your own data for publication in these directories be sure to use the recommended protocol to avoid confusion at a later date. Should you make a mistake, it will favor a lessor amount. You could get an inquiry from a prospective communications client asking if the AMSL value of 1105' is available on your tower. A simple calculation will tell you that your client wants to know if the 125' level on your tower is free (1105'-980' = 125').

In normal practice this 125' level is sometimes referred to as the radiation center, *rad center* and in reality would be the *exact center* of the antenna not its base mounting point. If your client wanted to use a 10' *stick*; 5' of the mast would extend *above* the 125' level to the 130' foot level on your tower; and 5' would extend below the 125' level to 120' level on your tower. The stick *mounting point* would then be at the 120' level extending to the *antenna tip* at the 130' level with the *rad center* at the desired 125' level.

If you are confused by this, draw a small to scale sketch of your tower and then draw the proposed mounting for the client's designated antenna. As can be seen, a ten foot antenna might prohibit your using a portion of your tower for another client not only at the mounting point but for several feet above and below the antenna due to interference or pattern distortion considerations. That is why it is important to also obtain the specifications for the proposed antenna and the operating frequency to be utilized in order to allow you to maximize your mounting points.

6.4 Site Identification - A BIG SIGN

The importance of a descriptive sign attached to your tower or security fence can not be over emphasized. If your access road prevents a site finder from driving to your site for a *look-see,* then also place a second sign at the road gate. On city buildings or other structures, a small sign by the elevator or a tasteful sign on the side of the building facing the heaviest traffic should be considered. The sign should prominently state **Antenna Sites Available**. Also give the AGL, the AMSL, the street address and the Latitude and Longitude in NAD83. This data does not have to be prominently displayed. Finally give the telephone number or e-mail address for obtaining more information.

Fig. 6.4 Dual purpose site sign

A basic sign can be provided by a local sign maker using the new computer/vinyl cutting system on a pre painted and drilled heavy gage aluminum plate. Mount the sign securely using screws or heavy gage wire. Position the sign above vegetation growth out in the open for the best visibility from a car window. Consider night lighting if feasible. If your tower is along a highway, post a large sign right on the tower leg for best visibility. Do not be afraid to advertise your site. You would be amazed at how many towers I've surveyed for possible clients that didn't have any identification on them what so ever. Not so much as a company name! If you want to rent your tower to communication people you have got to let them know it is available. Signs account for 50% of the first contact.

Another technique which has great potential for attracting engineering interest in your site is an *empty platform.* At the time your tower is constructed install an empty mounting platform or mounting bracket array with empty pipes on them and out in the open and as near to the top as you can go. See Figure 5.11.5. These hardware options represent about 4% or 5% of the overall cost of the tower. However it is a recoverable expense when adding a client to your tower. Because the platform is in place, installation for the client is minimized and the mounting system can be included in an initial attachment fee. It will actually save your client several weeks (to order the *steel*) and be less expensive because it was installed at the time the tower was being erected and not after market. Having a second or third platform in place also has the advantage of minimizing the possibility of damaging another client's facility upon installation. Place the platform in the next best position. It is usually what will be sought anyway. If not, the platform can easily be repositioned and reoriented while on the tower.

In the mean time you are showing everyone who considers your site that you have excess capacity ready for immediate use even without a sign. I once proposed painting a huge **red D** on company owned towers where I was previously employed. My company didn't think too much of the idea but I felt it would certainly get the message across to the right people. Well maybe it wasn't so great an idea after all. But anything that might work is worth a try and since it is *your* facility you can do as you please.

6.5 The Antenna Site Brochure

The first document required for your marketing arsenal is a site brochure. This document need not be anything other than an 8 ½" by 11" sheet of *white* 60lb. paper consisting of locating maps, (both *topographic* and if necessary a *road map*) showing the exact location and directions. Be sure to verify your written directions. Enlist the assistance of someone you know who can't help but get lost. It is not good enough to put an "x" on the map if the road to the tower is hidden behind some bushes or is a right-of-way across someone's property which might be hidden at the end of a private driveway. Be specific with distances and use clearly written terms. These directions will be used *over and over* again if your site is selected. It will be used by the consultants who must visit the site and make property determinations and title searches, by the technicians and installers who will be visiting from time to time during the construction phase, and by servicemen who will maintain the equipment. The map should also show the major service areas or secondary areas which can be served by your facility. Actually this information is what first attracts interest in your site over any other parameter.

A typical brochure should be clean, informative and *fax several times* without degradation. After you have come up with a preliminary design try faxing it *twice*. Use the first fax to send the second fax. More than likely you will loose some vital information in this process but you can quickly change the type face to one that holds up to the degradation of multiple faxing. Be sure to put the state, county, local municipal authority, location (township/city/borough) zoning and the target area served. I once identified a site in North Bergen, New Jersey, with little interest, until I later mentioned that it covered most of New York City within a ten (10) mile radius.

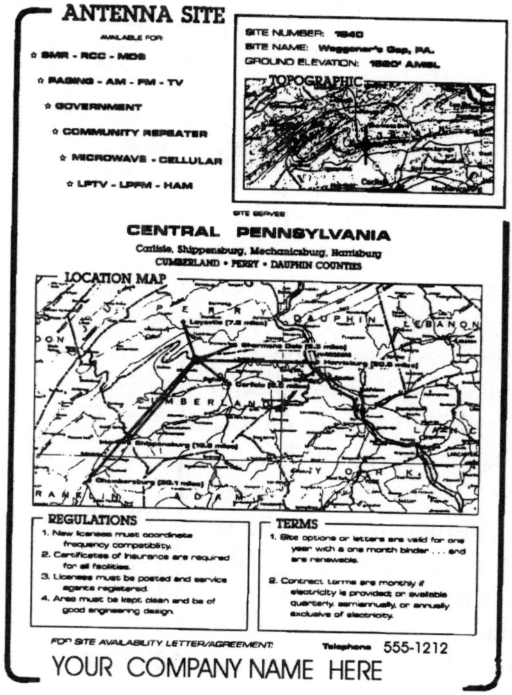

FIG. 6.51 TYPICAL SITE BROCHURE SIDE A

Include precise NAD83 Longitude and Latitude, ground elevations, HAAT, structure heights, telephone company (name), power company (name), FCC & FAA site registration numbers, nearest airport in miles and direction; a written description of the reliable coverage area and population in that area. If you are marketing a rural site or a corridor site try to list the portions of the major Interstates covered and the number of vehicles traveling over this road per day. This data is often available from your state's highway department. If you are covering a small town, what larger city does the small community support? Use the MSA number. Try to be specific.

Finally, since you know your own territory better than do the consultants, try to outline a known coverage problem in your area that your facility covers. We had a local cellular company build its MTSO near a pretty deep three mile dip in an four lane road leading out of town. They thought the site could handle this hill. Invariably portables *dropped calls* upon reaching the bottom of the hill. They had no choice but to build another tower within two miles of the original site. They used this site to solve another coverage problem. It did that ok but calls still dropped at the bottom of the original hill. A third tower was constructed to finally solve the problem. Very costly engineering by consultants not familiar with the territory. There wasn't a ham radio operator, CB'er or business band radio user who wasn't familiar with this hill. Most local radio technicians had solved the problem years ago. Unfortunately they did so without providing the necessary capacity to attract a cellular client, i.e. too light a tower on restricted land.

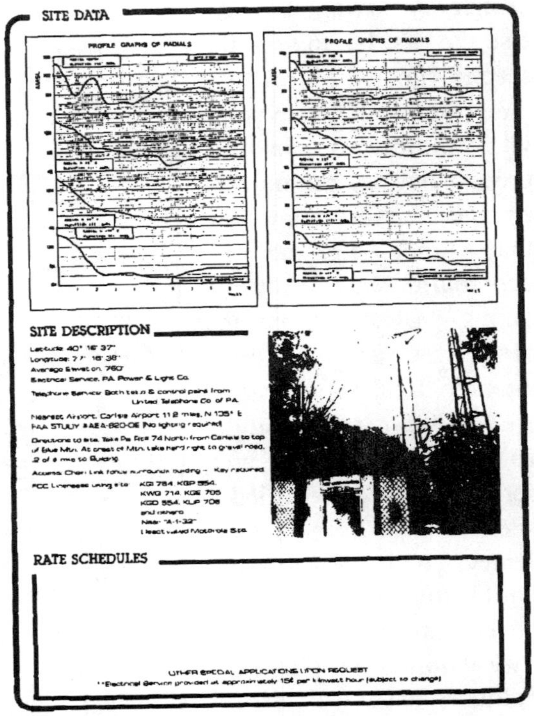

FIG. 6.52 PROMOTIONAL BROCHURE SIDE B

Photographs of your site, site views and buildings can also be effective. Good clear black and white photographs or laser printed pictures can convey an immediate impression of the professionalism of your facility. If your site cannot provide this professional image, don't use pictures right away. As the site develops you will eventually achieve the standard you will be proud to promote. Preparation of a self-contained slide show with text, on a 3 ½" floppy disk is also acceptable. This package should have the software viewer built into the images so that anyone can run an executable (.exe) program to see exactly what you want them to see and what you have to offer.

One piece of software capable of making your brochure and slide show, including an inexpensive scanner, is *Easy Photo Design Works & EasyPhoto* by Storm Technology. **info@easyphoto.com** This package is very easy to learn and use insuring an immediate product with nothing more than a few snapshots and some text notes. However there are many presentation packages and software applications which will more than adequately handle the job. Keep in mind also that your brochure should be able to be converted to **html** format for uploading to your ISP (Internet Service Provider). Often these services provide a free Web page in addition to your e-mail and Internet access. Using this method for the distribution of your site data can be another effective way to provide additional information.

6.6 Direct Mail Program

There are several data bases you must develop in order to implement an effective direct mail program. Your objective is to get your site information to design engineers, consultants, carriers and Site Acquisition Specialists. One source for effective mailing information are the trade journals which serve the industry. In particular, *RCR* a weekly newspaper **www.rcrnews.com** and *Wireless Week*, **www.wirelessweek.com** These two publications periodically profile top performers and actually give names addresses and titles along with markets served and telephone numbers. In addition to what is published from time to time they offer directories and reports which overview the entire industry. These reports vary in terms by publication. Sometimes they are summarized by Web site, others by state, by markets, by type of service, by personalities, by stock classification etc. Not only do trade publications provide necessary data you'll need they also provide a vehicle for you to advertise your own site information in classified sections.

Many major players also use institutional advertising in these publications not only providing Web addresses but corporate objectives and service offerings. Log on to Web pages identified, locate the engineering department and e-mail your site data directly to various people in the organization. Be sure to designate that it is an *antenna site offering* so that the folks you are sending your information to do not think its junk mail. A short e-mail followed by a neat colorful page presenting your site is your objective. Send it to everybody whose individual e-mail addresses are listed including the CEO and financial people. This will take you a couple of hours per month. But these sales targets represent the most current and effective mailing list you could possibly develop and at very little expense, i.e. no printing bills, postage or mailing costs. Hopefully, the person you e-mail will merely forward the data to the correct person in the organization or consultant.

6.7 The Management Company Connection

In addition to the actual communication companies you are attempting to target, management consultants also advertise. These folks represent another secondary market for your site promotion activities. When soliciting these companies it is important to let them know that *agents and brokers will be protected.* What this means is that if a communication company completely ignores your site and facilities but one of their consultants finds your site and proposes its use, the consultant serving as the *procuring cause* should be entitled to a commission if your site is selected.

The issue is do you want to share your site revenue with an agent or not get the business at all? This is a decision which you will need to make very early in your business plan. In order to do this you will have to calculate the impact of agent commissions vs. your own sales abilities to procure the business without the agent. Should you place the marketing functions exclusively with a tower representation company then a percentage of *all business* which comes to your tower will belong to the management company. This arrangement allows you to delegate the entire tower operation to what amounts to a property manager. There are distinct advantages in using this process. Depending upon the extent of delegation, you no longer have to even concern yourself with the day to day operations of your site. You would merely be paid for your investment. However if your objective is *hands on management*, you need not sign an *exclusive* management agreement with any one company.

Rather you would offer an *agency* to any management company who brings you business. This arrangement is very much like an *open listing* in real estate with the major difference that the property can be sold again and again and commissioning might be a continuing share of the rental revenue over a period of years. Any arrangement needs to be comfortable to all the parties. You are looking for management companies who will not take unfair advantage of your facility and who will be pro-active in the marketing your site. These companies are looking for site properties/owners who are cooperative, reasonable and whose facilities will meet their communication client's needs allowing them also to make a profit.

Ethical behavior on your part requires that you do not try to undercut sincere negotiating efforts by an agent with their client before a deal is signed. Unfortunately this situation is quite common in the industry and has created an unusual sequence of negotiating steps which require consideration.

Very soon after your facility has been completed you will find yourself also being solicited by tower or site management firms looking to represent your interests. Initial solicitations will be in the form of a letter or phone call introducing themselves, the clients they serve and listing the services they offer.

Typically such firms will review your site and give you recommendations as to how it can be improved to serve a certain class of subscriber. In addition they will promote your site in "their" data base and bring interested communications firms to you for site consideration.

If you have an exclusive site representation agreement, they will negotiate the site lease, commission rates and approve all engineering and mechanical construction proposals. They do this in order to maximize your facilities. In addition they can provide access management, billing and collection services, audit and periodic site inspections.

Management firms are very familiar with what is going on in your market. Competent firms rarely leave any money on the table. They are preferred by communication companies because there is no question once the rates and facilities have been established that agreements will be forthcoming and ready for execution within days, not weeks!

But sometimes it is a *love-hate* relationship. The reason is that a communication company may benefit from negotiations with an unsophisticated site owner directly and obtain a site at a considerable savings over a managed site. For this reason it is prudent for you to inventory your potential competition within your coverage area and adjust your rates and terms accordingly. Yes, you certainly can hold out for a $2000 per month site rent deal, but figuring that if an alternative piece of land can be leased for $300 per month and a tower constructed for $70,000, as a commissioned alternative for the consultants, your "20 year deal at $2000" just won't materialize.

Negotiating is the management consultant's specialty. But the *time game* is not one which you will want to play to exact a higher rent. The reason is that in almost every case in addition to your site, the communication company has an alternative, a *B site* (even a C, D and E site) which has been identified. These alternate sites can be brought into the picture quickly if negotiations break down with you. Site management firms know this strategy and use it effectively for everyone's benefit. But they need room for negotiations and if you are represented you must give it to them.

Commissions are negotiable but generally follow similar property management guidelines in the 20% to 35% range depending upon services provided. I hesitate to suggest any rate as all rates are negotiable, but it is important to have some basis for reference. I have seen rates of 50% commission where the site management company actually had an equity interest in a portion of the tower. I have also seen rates as low as 10% in situations where the management company performed minimally on behalf of the site owner. A typical management agreement attempts to set a fixed commission amount which is usually applied to the monthly rental. Some contracts provide this percentage for the duration of the communication company's use of the facility.

For example. If you charge a communications company $1000 per month for your site and the management company brought you the deal and is performing all the services at a 25% commission rate, they would be entitled to $250 per month. Your share would be $750.00 per month. Site owners often wonder if this *cut* is commensurate with the management service provided. *Why are they entitled to this amount for the entire duration of the contract*? In some cases it could be up to 25 years.

There are two ways of looking at this. The first is would you have obtained the account on your own? More importantly would you have been able to negotiate the price which you are receiving? In my experience the site owner either tends to under price or over price his facility. The key to a good deal is *proper pricing* based upon market conditions and use value. The theory is much the same when determining the price of a residential property. Real estate agents create a Comparative Market Analysis *CMA* by carefully analyzing what the market forces have to say about pricing the property. Sellers often feel that the carpet they installed fifteen years ago, or the deck that grandpa built by hand, adds value to the property which should be added into the price. Referred to as upgrades many of these amenities (and that is all they are) usually add nothing to the price. Whether a property is over valued or undervalued depends upon what a buyer is willing to pay.

The same holds true in the site business. If the site is 20% over priced, you might only get 20% of the business. If the price is 10% over priced you could expect 50% of the business and if it is priced correctly you can expect 95% of the business.

Another important consideration is cost analysis, not yours . . . but your clients. If you are the only tower within miles, and no others will be authorized, and your site serves a million potential subscribers per day, you can charge what ever you want! I actually know of many locations which really meet this criteria. As basic as it sounds *location* is the key. But in an otherwise imperfect world with competitors, relying upon marketing factors and good sense will insure you a good market share with a profit. A good practice is to *work the numbers.* Pretend **you** are the communication company and are faced with the decision to rent your site or build an alternative. If you can present good arguments for your site, you will get the business. The Ben Franklin close?

Now here is the other argument. When you are ready to market your tower or building you will be contacted by several management companies. You will hear this question: *Mr Site owner, if I can get you more than you are asking, in addition to my commission will you let me negotiate the contract and bring you a communications account?* Your answer depends upon many factors. Keep in mind that a 10% commission against a $1000 a month rental is only $100 per month. But a 25% commission against a $1500 a month rental is $375.00 per month. Your income on this transaction is $1125 per month. Which management company is going to work harder for you?

We are also setting you up a bit. No doubt if you are successful in marketing your site you will recognized other site opportunities in your area and can serve as a representative for these building and land owners as *their* management representative which adds *their* facilities to *your* portfolio. Once established your customer base might be interested in all the sites in your developing network. If you can attract investment capital so much the better. You would then be free to joint venture in any number of ways to develop additional sites. This process can be relatively risk free if you don't build anything until you have contracts to cover expenses. But of course the most obvious area for immediate development are existing buildings or structures. Perhaps one of your best sources for joint venturing in the area of new site development is your client. Often, after the initial deal is cut, you can go back and revisit the deal with the client making capital investments or planned expansions which meet both your objectives. These deals are done every day but are usually practical and less risky once the landlord/tenant relationship has been established.

Management companies don't usually involve themselves in these sorts of twists and turns and sometimes might even get in the way, so care should be taken to contractually reserve your right to change a client from a lessee to a partner. One way to do this is to provide for an optional buy-down of the commission with the management company. Should any conflict arise it is imperative that there be no doubt who was the procuring cause. If you believe that a firm should be paid for what they do or did, then you will have few problems in meeting your business plan using a management consultant.

6.8 The Site Management Company

There are many approaches to the representation agreement. Comparing them is relatively easy. You will receive a copy from any management company you contact probably by overnight express.

Alternately, call a few tower or site operators and ask them what they have and for a copy of their management agreement. You can't be too careful at this point so do your homework. If you find a particularly suitable agreement use it as a starting point. This will save you a lot of time and legal development fees. One caution, when using an attorney, be sure to either use a communication attorney or keep your discussions centered on contract law. There is no need to waste consultation time getting into communication issues. A similar area of law is real estate. You should be able to find many suitable specialists in this discipline without going too far from home. Upon reviewing other management agreements you will want to check several points.

1. What services will be provided on a continuing basis. Will the management company do the billing, collection, coordination of installation activities, foreclosure, assist you in the compliance with Federal Regulations (tower lighting, radiation monitoring and security), tower maintenance and inspection as well as a host of other items? Make a list. Some companies allow you to do all the site work and they are just the marketing agent. Others are turnkey services providing complete or custom tower management.

2. Check the fees. In most cases the compensation will be a percentage of the site rental agreement they bring you for the life of that site agreement. If they bring you a site rental agreement for five years (5) with three, (3) five-year renewals, they would be entitled to a percentage of your monthly rental receipts *as long as the site contract is in place.*

That's twenty years! In many instances if you change firms, say in year seven, the old firm will have wording in their agreement with you such that you must continue to pay them even after your agreement with them is terminated. This seems fair if the contract is lucrative, but your reason for changing firms might be for dissatisfaction, in which case continuing to pay them would be very unpleasant for you. From the management company perspective, they feel entitled to the deal as it was originally negotiated. The time to consider these issues is *before* the management company agreement is executed. In another instance, if your management company obtains or provides an equity interest in your site or facility, termination can be very difficult. This can and does happen so be careful to visualize the potential relationships which might evolve. Consult competent legal advice on the issues of concern to you.

3. Be sure to provide for a measure of performance standard which permits a de facto decision to terminate your management agreement without litigation. One very reputable company with whom I had experience promised in writing that if the management company failed to obtain a minimum of five percent (5%) increase per year during the life of the agreement it could be terminated. A pretty safe bet here. The way they accomplished this feat is to negotiate an **escalation clause** in site rental agreements of a similar amount. There are other concerns however. One site management company has a very lengthy management agreement which contains no specified term, no specified dissolution procedure and few applicable services one of which is a lengthy procedure for *seizing* communication equipment from your property and selling it in the event of a rent payment default.

Except for smaller companies falling upon hard times, larger communication companies usually demand a waiver of possession to title on equipment placed at your site. Often the equipment is financed by a lender who requires a first lien on it. This doesn't mean you can't litigate a claim, it means you'll have to *step in line* as a creditor. In this regard, companies that have applied for reorganization protection have a real good chance of bouncing back or being acquired by a stronger company. If you work with them they will often reestablish their relationship with you. If the management company starts to play hard ball with a failing communication company (because they have hundreds of other sites for *other tower clients* and stand to loose a lot of commissions cumulatively) you can be sure everyone will loose. Do you have control of this situation in your agreement? Remember that the frequencies do not go away in a business reorganization. The only thing that goes away is the poor management that got them into the pickle in the first place. Your facility is part of their capability to provide a service. High salaries will be cut, people laid off but your antenna site will be one of the last areas of consideration. Without it, a troubled company loses its ability to make money. Your site rental agreement could be declared voidable under a reorganization anyway so you might like to stand by the company helping it get back on its feet. Will your management agreement permit you to do this?

4. Who is the agent for whom? This is one of the really difficult situations which seems to crop up from time to time. Let us explore this issue briefly. You have a tower site with several subscribers. You elect to go with an exclusive site manager. They establish a fiduciary relationship with you and will want a percentage of your existing business which they did not procure for you. Do you want to give it to them? What percentage should they have? You can elect to identify only business moving forward or to negotiate a rate for *walk-in business* or exclude it all together. Walk-in business is a *site finder* who knocks on your door, never heard of your management company and wants to do a direct deal with you. Does your management contract permit this? How do you know that the *site finder* didn't obtain the information on your site from your management company's web site. Would it matter if he did and didn't tell you? What are his motivations?

Most agency relationships require that you refer the *site finder* to your management company. In some cases, in roof-top sites for instance there are few knowledgeable personnel who can make the decisions quickly so the management company services are preferred. But in your case you can make the decision and are the site owner so naturally the *site finder* is going to try to *do an end run around your agent*. It happens all the time. If your site management agreement is not exclusive are you free to cut your own deal? If it is an exclusive right to manage your site, you may not have this right.

Let us take this situation a little deeper. Now suppose that a site management firm has been hired by a communication company to find or develop a network of sites. Your site happens to be in the footprint. They contact you. Remember this company does not represent you. You tell them you are represented by another company and to contact them. After a time a deal emerges. It provides rent income for you, commissions for your management company and commissions for the other company. The communication company will do the deal with you but will express discontent with the *double dipping going on* by all the consultant companies. Now what?

If the rates are above market value the communication company will go to an alternative cheaper site. Everyone except the management company representing the communication firm loses the business. If you do not have this situation spelled out in your agreement, neither management company will want to budge in their position. The embarrassment comes from the fact that your direct agreement with the communication company contains a price consisting of three (3) components. What it won't contain is the actual entitlement provided to your agent. That would be in your separate management agreement. Also the company that is serving as the procuring cause wants some sort of *side deal* with you so they can pick you and get paid. They may not wish to disclose this deal to anyone. These deals can get very complicated and communication companies are beginning to question these agency relationships. They want to see the overhead costs involved.

<u>The Lease- Sub-Lease.</u> One way around the above situation is to change who the customer is. The management firm leases the site from you with the right to sublease. They then lease the site to the communication firm. Sounds pretty easy except that there are two rates involved for the same site. Yours is the wholesale lease, the management firm's lease is the retail price and the difference is their margin. Sounds good on paper except that default clauses become dependant upon both leases. Secondly, communication companies in an effort to standardize liability and procedure often request a plot plan and title insurance for your site. They then record the lease at the local courthouse. The first problem to arise before doing this is the question of *agency* for the management company writing its lease with communication company. Usually communication companies require assurances that the subagent has the unrestricted right to lease your property and in so doing an implied agency is required which eventually must be reduced to writing. In order to do this you now have an apparent conflicting agency relationship with a customer. The communication companies will more than likely try to cut out the middle man. Can you blame them? A contractual situation of this type can sometimes expand to four subagents. Two representing you and two representing the communication consultant. In a *feeding frenzy* for commissions, contracts can get pretty complicated. If things get too ridiculous, communication companies will stop calling and move on to their alternative site. You will have lost the business not to engineering considerations (which should be the only reason you should ever loose a

client) but to the dealings of the agents involved.

There is a way out of this dilemma. One proposal is to pay a one time finder's fee from your rent proceeds. You are buying down an agent with a specific set of terms *take it or leave it.* The communication company is not paying this commission, you are from your proceeds. This can be a one month payment in full, or several payments over a period of time depending upon the strength of the site agreement or other processing incentives. If you have this as an established policy prior to the deal, you will find that an agent who is persistently trying to get more from you by charging the client more will usually acquiesce. After all it is the way **you do** business. These commissions can be very lucrative and serve as an excellent incentive for unsolicited agents to give you the business. If the deal falls through with the agent and the communication client eventually contacts you directly, the price will be the same. You won't look like you are trying to take advantage of the situation. The bottom line is that you remain the reputable site provider and your reputation will quickly tell others of your fairness.

6.9 Other Management Programs

Once you have signed on with a management team you should continue to assist them in the performance of their job. They will be representing many other site owners like yourself, and they may even be building towers for their own portfolio. In addition, some relatively new concepts have emerged. Some investment opportunities called *Build-to-Suit* and *Purchase Leaseback* programs have cropped up all over the country. These programs essentially apply to towers and are less attractive or even applicable to existing non-tower structures such as buildings.

Build To Suit Tower Program. A communication company signs a deal with a tower management or investment firm. Then selects a series of sites where towers need to be constructed. The management company constructs the towers and rents them to the communication company for as long as it takes to pay them off. However the management company adds one more consideration. They construct the tower for multiple purpose use and proceed to promote additional business for this tower (or network of towers). The original user, the *anchor tenant,* gets an exact specification tower for their system. They may even negotiate lower start up rents i.e. rents that normally wouldn't cover the amortized payment of the facility for the investor until after a few years when the rent would go back up. Another variation on this is to charge the anchor tenant the full amortized amount for the tower until a second tenant is obtained. Then the anchor tenant's monthly payment is reduced.

Communication firms like these programs because it moves investment capitalization to someone else. Wall Street views the practice as good asset management. Other views suggest that new communication firms are willing to mortgage their future for the sake of start up efficiency. This allows the company to put more of its funding into the business of making money. However, like anything else there are some caveats. The communication company might not appreciate a second and competing company from just jumping on an existing tower which they have developed. That is why they usually like to negotiate a lower monthly rate when the second tenant arrives. The second tenant pays full price. Their own rent and the difference between the anchor tenant's rent and the amortization amount. Everything works if the investment tower company can attract multiple users or co-located facilities.

Build to suit usually involves *corridor* builds or building towers along highways. One way a network of towers can be frustrated would be if a series of towers in a build to suit program was established but the communication company found a few alternative existing structures along the way. This would mean that the tower company investor wouldn't have a complete system. There would be a *hole* in the system that they didn't control. This situation may give some leverage to the communication company in that it now knows your *system* is incomplete and less desirable to another client if you can't also control the alternatives.

Now supposing your tower is the alternative. You have a tower on your property along an Interstate highway. Your rents are competitive. Another communication company elects to employ a build to suit program along the same highway until reaching your tower. They want to cut a deal with you for two reasons. Your rates might be less expensive than the build to suit tower or they might intentionally want to frustrate a competitor from coming in and taking advantage of their developmental engineering and facilities planning. They could also build their own tower here to do the same thing. Of course this might suggest a slight adversarial position with their build to suit vendor. The tower management company looks at their system and sees the *hole.* They may elect to build a tower anyway just so they can have the entire network. Or they may cut a deal with you. They might even offer to buy you out. After all, you would be a competitor at this point.

In the early days of build to suit most projects were transactional. Open ended agreements, which were non-binding in nature, allowed both the investment company and the communication company to pick and choose their projects. Communication companies when faced with constructing a site for themselves would more than likely build a low cost single application facility. If however they could joint venture with another communication company, usually on a one for one basis, a multiple use facility would be constructed and shared. With the infusion of so much investment capital in long term and high payback tower projects communication companies now solicit competitively for build to suit vendors looking to receive exclusive territories for the investor in return for better and lower monthly rental rates. The popularity of these programs is wide spread. The money seems to be flowing into these programs without difficulty. The caution in this venture is if no other communication companies show up to use the towers. A paradoxical situation is also quite possible. What if all of the communication companies in a given market implement only a build to suit program such that there is no one left to be a co-location tenant? Investors beware it is happening as we speak.

Purchase/Leaseback Tower Program. Another variation of the build to suit program results from the sales efforts of enthusiastic tower management companies. Publicly financed corporations or subsidiaries acting as development specialists offer to buy the existing towers from communication companies for cash or terms and then rent them back to the same communication company. This process returns working capital to the communication company for other more pressing projects while at the same time reducing the company's debt-to-income ratio making it look a bit more attractive to investors. One problem immediately arises however if the communication company borrowed the money to build their own tower in the first place. In this case the asset has been pledged to the financial institution title to which can not be conveyed without a payoff. Another problem is that most existing towers are often low bid single application towers and will not easily support additional co-locators. In the event the actual tower location is in a controlled area or has been grand fathered such that improvements may be made but no new towers will be permitted, the purchase/leaseback program can be a good investment in spite of the augmentations which would have to be made in order to improve the tower's multiple loading capacity.

6.10 "Junk Sites"

There are management companies who specialize in soliciting as many siting opportunities as possible regardless of their quality. Although not necessarily under contract nor even under control, individual and potential antenna sites are nevertheless catalogued, identified and placed as an integrated package before communication companies for their considerations. A friend of mine in the paging business referred to these sites as *junk sites*. The problem is that full engineering must done to see if the location published is suitable. Upon visiting the actual site we might be looking at a telephone pole in someone's back yard.

In some instances potential sites are literally signed up as a result of just one direct mailing program. The brochure might read something like this *Cash in on your location by providing a lucrative antenna site for the expanding telecommunications industry*. You read the material, sign the agreement send it back, and you are then given the impression that your property is being marketed effectively. Doing this, some companies boast of thousands of sites *ready for immediate use* by communication companies. I once reviewed a web page for one similar offering and noticed that there were about thirty (30) sites in one small town of interest whose Longitude was the same but whose Latitude varied by only a couple of seconds from site to site. Upon inquiring I was told that the management company had *signed* a small shopping mall and every store location was thrown in the mix. In addition, all of the shops were one story only. The quality of these thirty (30) sites left much to be desired.

Be sure, if I had wanted to put an antenna on one of these shops I could most certainly have done so. But it is highly unlikely that would happen under the circumstances. What may be of further interest is that some firms might charge unsuspecting potential site owners *a nominal fee for registering or servicing their location in a site portfolio*. It is important to read management material carefully so that the *thousands of sites offered* are in fact valid antenna sites and that your valid site is not thrown in with *junk sites*. One easy way to verify this is to obtain a copy of the FCC data base on CD ROM for your state. Search the co-ordinates against those published. If there is any *licensed* activity at the site you select, it will more than likely be a good and desirable one. Incidently this is also a good way to track communication companies and see where they are presently located. Helpful information when attempting to determine what locations have the best potential for future development.

Finally, recognized site catalogs previously mentioned in this chapter are reputable directory publishers and should not be confused with site company solicitations of the type just described. The important point is that every effort should be made to check things out and verify what is actually going to happen *before* signing on the dotted line.

A Few Tower Management Firms

Apex Site Management	www.apexsiter.com
Hutton	www.huttoncom.com
Pinnacle Towers	www.pinnacletowers.com
WFI Wireless Facilities Inc.	1.800.935-6835
Metapath	www.metapath.com
CLS Group	www.capitalland.com
Unisite	www.unisite.com
TeleCom Towers LLC	1.800 559-1969
American Tower Corp.	www.americantower.com
SpectraSite	www.SpectraSitwe.com

CHAPTER SEVEN

7.1 The Business Structure of Site Ownership

Site ownership and operation can take many forms. It can be a sole proprietorship, general or limited partnership; several limited liability entities such as a limited liability partnership, C-corporation or a Sub-Chapter S Corporation. How you organize for business and where you take it depends upon the business laws of your state and other objectives. For example. What if you want to sell your site business someday? Obviously the simplest form of business entity is the sole proprietorship. In this organization everything belongs to you. You make all the decisions and can operate the site as you please. You may handle all the development, construction and marketing. You are responsible for income reporting and expenses as they occur. You may also contract for services as required. Since much of your original plant will be a capital investment, you may also be entitled to depreciation for your fixed assets such as a building or the tower.

The sole proprietorship, is the easiest to form, alter and liquidate and in most cases the entire reporting for this type of organization will fit nicely on the IRS *Schedule C* of the personal income tax 1040 return. What may not be immediately apparent is that some of your personal tools, labor and transportation might be allocated to the tower site venture and thus be permissible deductions under stringent criteria if appropriately applied. The down side to this type of organization is that it in addition to putting up all the money for construction and development (as opposed to a partner or partners) should anything go wrong *you* could be personally liable to the extent of your *personal* and real property in addition to the site venture.

If you contemplate the use of a roof-top under an existing business entity such as a property company, church or other business organization the use of a wholly owned subsidiary might provide some degree of separation from your normal operation. In many ventures it is just included as another rental opportunity under the umbrella of the main business which is used. Whatever the case it will be a lot easier to get started if the organization and accounting system have already been established.

Risk considerations are important to the type of business entity selected. What is the extent of your liability if a technician falls from your tower? Or how about a thrill seeker who vaults your fence and climbs it for fun? What if the tower falls over and crushes a vehicle or injuries someone? What if a technician working on your roof-top drops a screwdriver to the street below? The underlying presumption in all these questions is personal injury and liability. You must assume you will always be *liable indeed.* The protection which you afford yourself in your dealings with your clients, i. e. waiver's of liability, responsibility clauses and the like, go a long way toward assigning fault and negligence **and must be done**. But that will not protect you from being named in a blanket lawsuit. Adjudication alone, even if you are totally innocent and free of negligence can be quite a costly undertaking.

You might find that run of the mill liability and casualty insurance companies will be a bit skeptical of insuring your telecommunication tower or roof-top. Others however, who specialize in this area, are not difficult to find. These firms insure carriers. They understand the industry and its risk factors. They are capable of underwriting reasonably priced insurance policies for your business objectives. The question is do you want to insulate yourself from the loss of personal assets to the extent that your siting enterprise is protected?

This issue should be considered with competent insurance assistance but should not be a deterrent from your goal of providing a lucrative site business. As will be seen in several site agreement examples, often you as a landlord will require that you (and your business entity) be named as a *co-insured on a multi-million dollar liability policy* which must be effective on your site prior to your client even showing up. Most communication companies have no problem with this and the process of sharing the risk management is not foreign nor completely unacceptable to your customers. They have a lot more to loose than you do at this point. In a liability suit the injured person usually goes after the entity with the deepest pockets. They name everyone including you because you are *there*. And you must pay someone to defend you.

By establishing a C-corporation, a limited or general partnership (structures vary from state to state so consult your attorney or state business bureau for information and regulations) you may guard your personal property from risk to the extent of any limited liability status. However in a full corporation there are other costs and reporting requirements that may not be worth the effort. Earned income for example is taxed twice. A *foreign* corporation is another possibility. Incorporating in a state with less paperwork, you then register in the state in which you intend to operate doing so as a *foreign* corporation.

7.2 Fictitious Names

One of the really neat tasks you have is the naming of your enterprise. The easiest and least costly is to just use your name. Most state regulations will permit you to use your name without registering it as a fictitious name which can be another costly and lengthy process. Initials in some states are also permissible. I have a friend who has been in the real estate business for years. Her name is Joan *Spire,* a great name for a tower business. You must also check to see that no one else is using the name. This process is usually pretty simple and will not cost too much if done in only one state. But a problem arises if you intend to do business all over the United States. Under these circumstances you might have to clear and register the name *in all the states* you intend to do business. I am reminded of *American Tower Corporation Inc.* a Texas based company and *American Tower Systems Inc.* out of Massachusetts. Here we had two independent companies operating nationwide with names so close so as to cause some confusion. If the name you use is "Jones Tower Company" usually you will not have to apply for a fictitious name if your name is Jones. However if you choose a name such as "High Antenna Sites" or other names of this type, (one actual company operating in my area calls itself *Clearshot Communications Inc.*,) you will not only need a fictitious name registration but it may be a good idea to register and trade protect any corporate identification as well. The name game is sure a lot of fun. After you select your name, you will need business cards, stationery, invoices and a sexy logo. You gotta have a sexy logo! These items can be very costly if delegated to an advertising agency so if you have any creativity at all now is the time to use it. Copyright everything and **be original**. A good image helps to attract attention. If you intend to sell your business in the future, it is a lot easer to sell a *name* other than your own.

7.3 The Need For "Clear Titles"

Most large communication companies will require a property description as an attachment to their rental agreement. They will be installing thousands of dollars worth of equipment at your site and need to be assured that you *indeed* own or control the land and that you have the right to lease it to them. Should you be a sole proprietorship, they will also require a succession of the right to lease to your heirs and assigns. If you die leaving your interest to a spouse or partner, they might want to tear down the tower and build a condo or something, Your estate (antenna site) might then be in breech of contract. But title troubles pop up all the time. Thus don't be surprised if the communication company performs a title search and obtains title insurance. What this means is that a third party title search is made against your property then insured naming the communication company as the beneficiary for stated *defects or clouds*. They pay for this not you. But your land and deed is what is being investigated. Hidden hazzards are the most dangerous problems that crop up in titles. Your abstract (the short history of ownership) might be perfect but your title could be worthless. Title insurance protects you from only some of these possibilities. So before you dig any holes on your land or land that you have leased, be sure to obtain a title search of your own as a condition to leasing the land or purchasing it. If you have owned the land for a while, it is also a good idea to *bring your title down to the present* by obtaining title insurance. Here are some common *problems* that could effect your right to use the land. False personation of the true owner of the land. Forged deeds.Instruments executed under expired or fabricated power of attorney. Delivered deeds after death of grantor or grantee, or without consent of the grantor. Deeds to or from defunct corporations. Undisclosed or missing heirs in a will of apparent intestate. Deed delivered by an unsound mind. Deed delivered by a minor or by aliens, or by people secretly married or unadopted persons or siblings born or adopted *after* a will was executed. Mistakes in recording the document. Errors in indexing. Deed delivered in lieu of foreclosure. *Ultra vires* (beyond or outside of authority) deeds given under false corporate resolution. Easements by prescription not discovered by a survey (a squatter has been living on your land for over 21 years and you have done nothing about it). Deed from a bigamous couple. A Federal condemnation without a filing of notice. Tax liens not paid. Undisclosed divorce of spouse who conveys as consort's heir. Marital rights of a spouse purportedly but not legally divorced. Any duress or intimidation in the execution of the deed instruments. Easements and right-of way disclosures. Zoning changes restricting the use of the land.

Some of the more common items frequently occurring in the title game can happen and must be remedied prior to a clear title being issued. In some cases court petitions, appearances and decisions are the only way to clear titles if previous parties to the deed are long gone and can't be found. A good title search and title insurance is your first line of defense in this type of guarantee. It is a task you will need to perform to insure the availability of your site. If you have already done the search and have insurance, you can refer your communication client to the title company *you have used* and they can bring down the title at a substantially reduced cost to the communication carrier. Sometimes however the communication carrier would rather use their own company. At any rate it is a requirement with which you will have to contend. It is always wise not to build anything until a *fee simple (absolute)* deed and clear title to the property is in front of you.

7.4 Agreement to Lease Real Property

If you do not own the land upon which you are going to construct a tower you must have a lease which reflects the nature of your intention. Typically such a lease, designed specifically for antenna siting purposes, is presented as a suggestion of items to consider:

THIS AGREEMENT TO LEASE REAL ESTATE (this "Agreement") is made as of the____day, of month, of year, between____(hereinafter "Lessor") and_____(hereinafter "Lessee") (collectively the "Parties").

WHEREAS, Lessor is the owner in fee simple of certain real property in (give location), more fully described as Exhibit A, and incorporated herein by reference (the "Property"); and

WHEREAS, Lessee desires to lease a sub part of the Property from Lessor, and Lessor desires to lease a sub part of the Property to Lessee on the terms and conditions set forth herein:

NOW THEREFORE, in consideration of the mutual promises contained herein and for other good and valuable consideration, receipt and sufficiency of which is hereby acknowledged, and in consideration of the terms and conditions set forth herein and with the intention of being legally bound hereby, Lessee and Lessor agree as follows:

The above text is so common that it can be found in one form or another in any real estate manual or transaction guide. Certain wordings may or may not be useable from state to state depending upon what laws have been enacted for the protection of parties to an agreement. Therefore it is strongly recommended that you check with your state's attorney general as to the suitability of such agreement and work with a competent attorney familiar with real estate leases. Many title companies have attorneys on staff who specialize in such leases by virtue of their work in real estate and thus can be a good starting point for your inquiries.

Moving into the detail of the lease paragraph by paragraph with some comments:

1. <u>LEASE:</u> Owner leases to Lessee portions of the Property (the "Leased Premises"), together with appurtenant Lease easements for ingress and egress, 24 hours per day, (*Very important or communication companies will not be interested*) seven days per week, on foot or by motor vehicle (including by truck), and to install and maintain transmission and utility wires and cables, conduits and pipes on the Property including over,

under or along a 20 foot right of way extending from the nearest public right-of-way for use by the public utility, at no cost to Lessee. (*This allows you to route power and telephone lines over other portions of the property without having to pay for that too*). In such event a new survey is required on the Property for the purposes of this Agreement, the cost of such survey shall be at the sole expense of the Lessee. (*You could try "Lessor" if you want to negotiate*).

(a) Term and Commencement of Lease: This Agreement shall commence upon the date upon which all zoning and permits necessary for Lessee to erect a _____foot tower (the "Tower") on the Property in a location as outlined in Exhibit B have been secured and Lessee commences such construction and erection (the "Commencement Date"), and shall expire on the twenty-fifth anniversary of the Commencement Date. If Lessee is unable to secure all required zoning and permits for the Tower within 180 days of the date upon which this Agreement is executed, either party may terminate this Agreement at its option with no further liability to the other party. *This section protects you if you can't build the tower.*

(b) Renewal: This Agreement shall automatically renew for four (4) successive five (5) year periods at a lease rate equal to the rate being paid at the conclusion of the prior term, plus inclusion of the annual adjustment described in paragraph 1 (c) (2) below, unless written notice from Lessee is provided to Lessor ninety (90) days prior to the termination of the then current term in effect.

(c) Lease Payments: Lessee shall pay Lessor payments as follows:

(1) Payment: Commencing upon the Commencement Date Lessee shall pay Lessor the sum of _(*written amount*)__ ($_____) monthly for the first twelve (12) months of the Agreement, *plus applicable sales tax*, which payments shall be payable to Lessor by the 15th. calendar day of each month.

(2) Annual Adjustment: Lessee and Lessor shall adjust

the monthly lease payment on the annual anniversary of the Commencement Date, and thereafter on every annual anniversary, increasing said amount by (*percent written*) (___%). *It is important to lock in any escalations at the beginning of the agreement. If you leave this open you won't know what to pass through to your clients and your land owner might allow you a few years to build up a nice customer base then sock it to you! Keep in mind he could still do it at the end of twenty-five years.*

2. LESSEE'S DEFAULT: In the event default by Lessee including without limitation nonpayment of any payment remains unpaid for thirty (30) Days from the date on which the Lessor gives Lessee written notice that said payment was due, Lessor may declare Lessee in default and may seek to evict Lessee. In the event of any litigation over the performance of this Agreement by either party, the prevailing party shall collect its reasonable costs and legal fees from the party which did not prevail. *Be careful with this clause. As a matter of routine the Lessee could withhold any payment, waiting until he got his "notice", then wait another 30 days to pay. It could become the routine rather than the intended procedure. I don't like this clause and would prefer liquidated damages accruing after a due date with an automatic drop dead default date. Think this through as it works against or for the Lessor or the Lessee depending who you are. One other point to keep in mind. your communication client may require some protection or intervention in the case of an eviction.*

3. LESSOR'S DEFAULT: In the event of default by Lessor hereunder, Lessee may, at its option, elect to enforce the terms hereof in any court of competent jurisdiction or demand and be entitled to any and all appropriate damages in connection with such default. *Some communication companies may try to locate the court in their home office area or state. See paragraph 15.*

4. TITLE: Lessor represents and warrants to Lessee that as of the signing of this Agreement it holds an unencumbered and marketable title to the Property and will provide satisfactory evidence thereof on or before the Commencement Date. If Lessor is unable to demonstrate on or before the Commencement Date that it has an unencumbered and marketable title to the Property, Lessor shall have thirty (30) days to remedy the title defects. In the event

that the defects cannot be remedied, Lessee shall have the right to cure and remedy the defect(s) and Lessor shall be liable for all damages and costs incurred by Lessee arising from the breach of the warranty concerning title. *See Section 7.3*

5. <u>POSSESSION:</u> Subject to paragraph 4 above, Lessor shall deliver and Lessee shall accept possession of the Property on the Commencement Date. Lessor represents that as of the Commencement Date, no party will possess any leasehold rights in or to the Property. Lessor shall have the right to continue to occupy the Property as well as the Lessee, and shall carry on such activities as are consistent with law and which do not interfere with Lessee's right to use the Property for a transmission facility as detailed in paragraph 8 (a) below. Lessor agrees that it shall not use the Property as of the signing of this Agreement except upon Lessee's prior consent, which consent shall not be withheld unreasonably. *A non-compete clause.*

6. <u>TAXES:</u> All taxes and assessments which are due and owing or have become a lien upon the Property as of the Commencement Date shall be paid by Lessor. Lessee shall have the right to cure and remedy any future unpaid and delinquent taxes or liens owing upon the Property and subtract a like amount, along with any reasonable costs and fees arising from the processing of such payments, from any future Lease payments payable to Lessor. *You are reminded that the Property as defined in this lease agreement could only be an agreed upon subpart of a greater track of land against which a tax lien would be placed. The Property as defined in the Exhibit A is not an authorized "sub-division" only a defined portion of the Lessor's land.*

7. <u>BROKER/AGENT PARTICIPATION:</u> Lessee and Lessor agree that no real estate brokerage or agent has participated in this transaction or otherwise has any claim for a commission resulting from the real estate lease agreed to herein. *Should the lease be coordinated by a management consultant or real estate agent then add them in and define their commission to the exclusion of others.*

8. <u>GENERAL PROVISION:</u> The following general provisions shall apply to the leasehold granted by this Agreement:

(a) <u>Purpose:</u> Lessee shall have the exclusive use of the

Property for purposes of operating communication systems, erecting a transmission tower, installing satellite antennas, transmitters and receivers, installing buildings or shelters and related and associated equipment both for the Lessee and its future clients. [See 8 (b)] Lessor will grant approval to the construction plans of Lessee prior to the Commencement Date. In the event that Lessor does not grant approval for the Lessee's plans, Lessee may at its option, declare this lease null and void. Lessee's operation of its facility shall at all times be consistent with law, ordinance and other applicable regulations. Lessor's initiation of Exhibit B shall serve as approval by Lessor of the location of the Leased Premises and appurtenant Lease easements which lessor also grants to Lessee. These include rights to (i) install electric service (lines) from public utility mains to its equipment; and, telephone service (lines) from public utility mains to its equipment. In addition (ii) to erect, construct or make Property improvements, alterations or additions ("Lessee's Changes") appropriate for Lessee's use. Lessee's Changes may include, as permitted by Law, fencing, roads and landscaping of the leased Premises. All Lessee's Changes shall remain Lessor's property.

(b) Assignment and Sublease: Lessee may freely assign its rights under this Agreement or sublet the Property, portions thereof or facilities on the property. *Without this provision there should be no deal. Some land owners will use this provision to attempt a percentage of your action with your clients. This is a logical substitution for lower monthly rents and Annual Adjustments. But watch out, it begins to take on elements of a joint venture and may give unwarranted control to your landlord.*

(c) Condition of Premises: Lessor agrees that no representation as to condition or repair of the Property, and no promise to alter, repair or improve the Property has been made nor implied by the Lessor which has not been so stated herein. At the expiration, Lessee shall deliver the Property to Lessor in good repair and shall remove all fixtures which it has placed **on** the property prior to expiration of the lease term, yielding up and delivering the

Property in like condition as when taken. *Be careful here. The antenna base and poured slabs for the shelters could be costly to remove. The key to this paragraph is the technicality of what is "on" the land not what is "in" the land. My recommendation is to try to exclude foundations from this paragraph such as "shall remove all fixtures" (except for poured foundations).*

(d) Right to Enter Property: Lessor shall have no right, except as may be required by law, to inspect or gain access to the leasehold improvements made by Lessee in or to the property. *This presumes that all leaseholds are the property of the Lessee and its clients. It is an FCC regulation that some access to equipment be restricted. If however anything belongs to the Lessor this right might not be granted. An example: Your shelter is wired to an automatic alarm which calls the local fire department. It "triggers" from a burned up transformer in one of your client's transmitters and the riot squad shows up on the Lessor's property. You are in Florida. You can bet your landlord is going to break into your building and turn the damn thing off. But on the other hand a nosey landlord is not entitled to snoop into your stuff. The solution is somewhere in between.*

(e) Alterations: Lessee may make alterations, additions or improvements in or to the Property, provided that said alterations, additions or improvements are compliant with federal, state and local laws, regulations or ordinances. Should any lien (mechanic's) be filed against the Property for work claimed to have been done on behalf of the Lessee, Lessee shall have fifteen (15) to correct such lien. Lessor shall have the right, but not the obligation, to pay or discharge any lien. If Lessor should elect to exercise this right, Lessee shall pay Lessor the amount plus any reasonable expenses on the next regular monthly rental due date. Lessor will allow Lessee to make reasonable and appropriate alterations to the Property in order to accomplish Lessee's changes, in accordance with plans and specifications ("Plans") to be submitted to Lessor to be approved not later than ten (10) days from submission. After Lessor's (i) acceptance of any Plans or, (ii) failure to respond in writing to Lessee's proposed Plans within ten (10) days of receipt or, (iii) failure to provide a written response within ten (10) days receipt of Plans revised by Lessee after comment from Lessor, the Plans shall be deemed accepted. After acceptance, Plans will be considered incorporated in the Lease as Exhibit "C".

(f) <u>Lessor is held harmless:</u> Lessor shall not be responsible nor liable to Lessee for any loss or damage that may occur by or through the acts or omissions of any one occupying any property adjacent to the Property. In addition, Lessor shall not be liable for any damage or injury to any person or property which occurs on the Property due to Lessee's or Lessee client's actions or omissions. Lessee shall maintain commercial liability insurance on the Property of a type and level consistent with good communications industry practice, and Lessor shall be named as an additional insured on said policy at no cost to the Lessor. A current Certificate of Insurance attesting to the above shall be provided by the Lessee to the Lessor and remain on file with the Lessor at all times during the term of this Lease. *See (7.1)*

(g) <u>Eminent Domain</u>: If during the term of this Agreement, proceedings shall be instituted under the power of eminent domain which shall be directed towards an eviction, total or partial, of Lessee, at the time the trial of such proceedings shall commence, this Agreement shall be voidable at the option of the Lessee and the terms above shall cease and terminate if Lessee elects to void this Agreement. If the lessee continues in possession, it shall be as a Lessee from month to month and for no longer terms, anything in this instrument to the contrary not withstanding.

(h) <u>Non Exclusive Remedies</u>: Each of the rights provided in this Agreement shall be cumulative.

(i) <u>No Right to Mortgage</u>: Lessor shall not have the right to subordinate this Agreement to any mortgage hereafter placed on the Property.

(j) Quiet Enjoyment: Upon Lessee paying the rent and performing all of the other provisions of this Agreement, Lessor agrees that Lessee shall peacefully and quietly have, hold and enjoy the Property during the term of this Agreement.

(l) Miscellaneous: The following additional terms and conditions apply to the performance of this Agreement by the parties:

(1) Lessor agrees that Lessee's ability to use the Leased Premises is contingent upon its suitability for Lessee's intended use from a technical engineering basis and Lessee's ability to obtain any and all governmental leases, permits, and approvals required or deemed necessary or appropriate by Lessee for its use of the Leased Premises (called "Governmental Approvals"). Lessor agrees to cooperate with Lessee, and at Lessee's expense join in any applications for Governmental Approvals. Lessor specifically authorizes Lessee to prepare, execute and file all necessary or appropriate applications to obtain Government Approvals for its use under this Lease.

(2) Lessor also agrees that Lessee's ability to use the Leased Premises is contingent upon the completion of contract arraignments with its Sub-licensees.

9. WAIVER OF DEFAULT: Either party may waive or remedy any default in any manner without such default remedied and without waiving any other prior or subsequent default; and wither party may waive or delay the exercise of any right or remedy under this Agreement without waiving that right or remedy or any other right or remedy hereunder. No waiver of any of the terms provisions or conditions hereof shall be effective against either party unless the same is in writing and signed by the waiving party.

10. <u>BINDING NATURE:</u> This Agreement shall be binding upon, and shall inure to the benefit of the respective heirs, executors, administrators, successors and assigns of the Parties.

11. <u>COVENANTS AND WARRANTIES</u>: Each of the agreements, covenants and warranties on the part of the Parties to this Agreement shall be deemed and construed to be on a continuing basis and shall survive the execution and delivery of this Agreement.

12. <u>NOTICES</u>: All notices, demands or requests required or permitted under this Agreement to be given by or to either of the Parties (i) shall be in writing and (ii) unless and until otherwise specified in a written notice by the respective parties or any of them, shall be sent to the parties at the following respective addresses:

If to the Lessor: (Name & Address)

If to the Lessee: (Name & Address)

or to such other address as either party from time to time shall designate by written notice to the other. Each such notice, demand or request shall be deemed to have been properly served for all purposes if personally delivered or sent by Registered or Certified Mail, return receipt requested, postage prepaid, to its addressee at its address as set forth herein above, or by recognized overnight courier service. Each such notice, demand or request so mailed by either party shall be deemed to have been received by its addressee on the first business day after the day of mailing.

13. <u>ENTIRE AGREEMENT</u>: This Agreement represents the entire agreement of the Parties as to the subject, and there exists no other oral or written agreement or understanding which is not a part of this Agreement or which shall have no force or effect. This Agreement shall not be amended or modified except by written instrument executed by the Parties.

14. SEPARABILITY: Each section, subsection and lesser section of this Agreement constitutes a separate and distinct undertaking, covenant or provision hereof. In the event that any provision of this Agreement shall be determined by a court of competent jurisdiction to be unlawful, such provision shall be deemed severed

from this Agreement, but every other provision of this Agreement shall remain in full force and effect.

15. GOVERNING LAWS: This Agreement shall be construed in accordance with and governed by the laws of (insert State). The Parties also agree that this Agreement was executed in (County) and that it is expected that its performance will remain in (County). Thus, the Parties agree that the sole venues for all actions brought pursuant to this Agreement shall be the courts of law or equity of (County) or in the courts of the United States of America located in (State).

16. TITLES & NAMES: Titles and headings at the beginning of paragraphs of this Agreement are used for the convenience of the reader and are of no force or effect. Pronouns, regardless of gender, shall be interpreted as if they were of the gender of the noun to which they refer.

17. INTERPRETATION: Neither this Agreement nor any provision contained herein shall be interpreted for or against either Party solely because that Party or that Party's legal representative prepared the provision.

18. COUNTERPARTS: This Agreement may be executed in one or more counterparts, each of which shall be deemed an original and all of which constitute the same Agreement.

IN WITNESS WHEREOF, the Parties hereto, intending to be legally bound, have executed and delivered this Agreement on the day and year first above written.

By____________________ Witness:

By____________________ Witness:

The purpose for presenting a site plot rental agreement in detail is to provide a basis for such a situation to the exclusion of other more traditional but similar raw land rental and leasing agreements. As can be seen, this agreement is specialized and simple enough to do the job when it is necessary to obtain the rights to land which you do not own. It must protect your capital investment and your business plan of sub-licensing portions of your facilities in any manner you choose to your communication customers.

Once the zoning process starts you may find your landlord changing his mind about your enterprise. I have seen some pretty naughty pressure applied to a land owner who quite innocently rented a portion of his land to a communication company to build a tower. Increased assessments, retribution, bad press releases and down right intimidation to influence this land owner to change his mind. Some call this process tortuous interference. Other's call it politics. It is not a bad idea to discuss the possibility up front in negotiations if you believe it will rear its ugly head.

Even if the land belongs to you personally, such an agreement will help define the isolated portion of your existing land by assigning the lease to your business entity. It establishes a cost center and provides other advantages in the allocation of revenue. In this case, an agreement would be between you as the land owner and your tower company. It provides a *virtual sub-division* when such authority may not be possible in your municipality. At any rate, the use of this specific agreement is only a suggestion and should be carefully reviewed for applicability to your given situation. The important point to remember is that more conventional land leases can be long, involved and **not allow** you to engage in the specialized business which is your intention.

7.5 License Agreement Background

Long ago and far away, it is told, two people who wished to agree on the exchange of real estate would gather up a bundle of twigs from the property and exchange them with a hand shake. The bundle of sticks supposedly has become a *bundle of rights* from then to now and has been replaced by tens of thousands of law books, statutes and attorneys. The written License agreement for space at your antenna site is the end result of these centuries of evolution in contract law. So much for Basic Law 101.

Oral agreements have their legal place in the world and can still be enforced with difficulty if a breech occurs. However as the agreement becomes more complex, writing things down and acknowledging what has been written as a true and unambiguous understanding makes a lot more sense. In most states, via their statues of fraud, particular contracts involving real property, must be in writing otherwise they are voidable. Most laws don't usually specify a strict requirement for a long legal tome using *whereas* and *therefore*. In most cases a simple memorandum or letter of agreement containing the description of the terms and conditions as agreed to by the parties will suffice.

But as can be seen by the description of the *termination clause*, in Section 7.6, which follows, *understanding* can travel in many subtle directions. That is why some of these paragraphs go to great lengths to describe as precisely and as clearly as possible, the point that is being negotiated. Sometimes this *preciseness* is filled with ambiguity and in some instances intentionally, unfortunate to say. So read the fine print (unless it is yours) and think about it from their perspective.

A contract is an agreement between parties for consideration (something of value). A contract without consideration is usually not a contract. The two most important components of a contract are the *offer* and the *acceptance*. One party makes an offer and the other party accepts the offer; or, counters with another offer in which case the offeree becomes the offeror and the process is reversed until there is a *meeting of the minds*.

Accordingly there must then be an offer, an acceptance and consideration in order to have an enforceable contract. When the contract is acted upon, it becomes much stronger. What this means is that if you are paid some rent by your tenant who begins construction of his facilities at your site, there is little doubt that the contract exists and that it is being validated through action.

Therefore it is obvious that the most important agreement you will need is the one you will use to sign-up your clients to install their equipment at your facilities. Note also that clients who will be using your facilities were referred to as *sub-licensee's* in the land agreement presented in Section 7.4. The terms *lease* and *license* need some explanation as they are widely used in this process and are sometimes confusing. Typically you (and your antenna site) are the *Licensor*. Your client would be the *Licensee*. If your client or *Licensee* is going to allow *another* entity to use their facilities at your antenna site, the third party might be called an *Assignee*. Wow . . .what is going on here?

Here is the problem. You have a rental agreement for the land, what amounts to a land lease. You know you want to sublease it (because you have given yourself the right to do that) to someone else, a communication company. However you have added value (a tower, building and power etc.) to the leased property. Now when you think about it, it might not be your intention to sublease the land at all (even though you might want to do *that* as well some time in the future) but rather to rent out the *use* of the land and your facilities as a *service*, "an antenna site service". If you are going to do this, then in effect you are granting authority or a *License,* i.e. a right granted which gives the communication company permission to do something which it could not legally do absent such permission. Because a license represents only a personal right, it is generally not assignable. In many instances antenna site agreements are referred to as LICENSE AGREEMENTS. Confusion arises because the communication companies are also *licensees* of communications frequencies by the Federal Communications Commission having also issued them *licenses* to operate their systems.

The nature and scope of these License Agreements vary from company to company. Smaller communication firms will trust you to supply the agreement similar to the example to follow. Larger companies will have their own agreements and will insist that it be the one that is used. Well, maybe! Some of these companies will play hard ball on this point but if your agreement covers the bases and is what you insist upon using in accordance with your policies, (how you treat everyone fairly and alike) there is often much room for compromise.

There are some basics that are usually not negotiable however. For one thing companies will insist upon a copy of the deed, will do a title search and will obtain title insurance. They will insist upon unencumbered twenty-four (24) hour access to their equipment. They will insist upon an FAA tower or roof-top registration number (an FAA study). They will insist upon full disclosure of agency relationships (your landlord). And they will insist upon recording their License Agreement with you at the local county courthouse. We do not want them to do that because our negotiated rates and special terms will now be public record and could be used against us by our competition. What we will allow is a single page memo indicating that a License Agreement has been executed between the parties but with no pricing data or details. In the event you are renting a roof-top the actual License Agreement could be attached to your deed as an obligation. We don't need our tenant having this power over our holdings. Their need to record is only to prove the existence of an arrangement should you be abducted by a flying saucer. They do not need to attach a lien on your property for a rental agreement. Be sure to resolve this issue in the agreement. They may not tell you they are recording it.

Optional items include a plot plan, a Phase I Environmental Study, perhaps a tower loading study, certainly a list of all the frequencies and antenna locations of your other tenants (to predict any intermodulation interference), some sort of reference to the new requirements for Non-Ionizing Radiation Rules and Regulations, and most importantly the all important Termination Clause.

Site Savvy

> Whether you use your own agreement or sign on with the one which has been provided by the communication company, the facilities to be installed should be specifically described.

In reviewing hundreds of tower site license agreements, I have seen efficient, binding and easy to understand agreements of no more than two (2) pages. I have also seen a license agreement by a major company that was over twenty-five (25) pages long and required two (2) witnesses to each signature including a notarized certification that the folks signing it had the authority to sign in the first place. Whether you use your own agreement or sign on with the one which has been provided by the communication company the facilities to be installed must be specifically set forth and defined, no more no less. In addition, the question of who supplies electrical power must be resolved; the waiver of liability and co-insured status should be established; when you are going to get paid; how much; for how long and what happens if you don't get paid. Once an agreement has been negotiated, and keep in mind every contract is negotiable to some extent, it won't change for almost twenty-five (25) years. Corporate attorneys sometimes get enamored with their obligation to protect their client against all things foreseen and unforseen thus making a relatively simple deal very complex. In the heart of an impasse I have also seen top management literally order their attorneys to make the deal work. I have also seen two separate interpretations of the same paragraph by two different attorneys working for the same communication company who authored the paragraph in the first place! The longer the agreement however, in my experience, the more there seems to be working against the site owner. So read the long ones carefully. It is important to note that communication firms desperately need what you can provide and all you need is a reasonable return on your investment. If these two conditions are met, it can truly be a *quid pro quo* relationship (something desired for something desired) working to everyone's advantage.

In addition, the speed at which negotiations are brought to fruition is also very important. Once your agreement is signed, even if the commencement date is a month or two away, the communication company dispenses with its alternative sites and moves on to the next problem, locking you into the system. So if you are offered a site deal, complete the negotiations and execution as quickly as possible. Do not refer it to an attorney who will sit on it for a few weeks. That could be the kiss of death. If you are prompt in your response you will not only get the deal, but you will assume a reputation for being cooperative. The best deals are ones where everybody leaves the table happy.

A word of caution. From time to time you may find that communication attorney's keep reworking the agreement in order to buy time to either check out an alternative site or because they are not ready to construct at your site for a while but don't want to loose the deal. They might string you along making frequent and minuscule changes to the agreement. If you suspect this is happening suggest an accommodation in a signed contract which will allow them to sign but not commence construction until they are more comfortable.

7.6 The *Termination Clause*

Now, lets look at the *termination clause*. Buried deep in the agreement where you are least likely to see it, is the *termination clause* which may read something like this:

Right to Terminate. Licensee may terminate this Agreement at its option, after giving not less than thirty days (30) days notice to Licensor, if: *(make it written notification not less than ninety days)*

(1) any government agency denies a (*pending*) request by the Licensee for, or revokes a permit, license or approval *(for reasons other than failure to construct on time*) which is required for Licensee to construct or operate any telecommunications facility on the premises;

(2) Licensee determines *(as evidenced by a competent engineering study suppled to the Licensor)* that *(external)* technical problems or radio interference from other existing antennas at the Premises *(at the time of the lease commencement date);* or interference received by nearby radio transmitting facilities, any of which create problems which cannot reasonably be corrected *(by either the*

Licensee or Licensor) and as such will clearly preclude Licensee from using the Premises for its intended purpose. *If you don't think this can happen it does. Here is an example. You have a one-hundred foot (100') tower with a communication company microwave dish at the top. Down the road, not on your property, a Tulip-popular tree after a few years, finally extends in height so as to block the microwave path. The little old lady who owns the tree won't allow you to trim the branches. Unless you can add some more height to your tower the game is over!*

(3) Licensee determines that Licensee does not have acceptable and legally enforceable means of ingress and egress to and from the Premises; *Sounds simple enough doesn't it? Remember this was one of the non-negotiable conditions mentioned earlier. Here is an example. An accounting firm occupied the top two floors of a tall building. The only access to the roof was from the top floor elevator to the final staircase to the penthouse. The accounting firm would not allow anyone to walk through their reception area during non business hours even going so far as installing a "lock" on the elevator denying access after business hours. The building owner had no recourse because his lease with the accounting firm contained no "right" for another tenant to access the roof through the area controlled by the firm. Big contract down the tubes.*

(4) Utilities necessary for Licensee's use of the Premises are not available to the Premises; or

(5) The Premises are damaged or destroyed to an extent which prohibits or materially interferes with Licensees's use of the Premises.

In the event of termination by Licensee pursuant to this provision, Licensee shall be relieved of all further liability hereunder, except its obligation to remove its improvements *(equipment)* as provided herein. Any rental fees paid prior to said termination date shall be retained by Licensor.

The above termination clause is about average. Now here are actual clauses proposed by well known communication firms which were <u>definitely unacceptable</u> to the site owner:

"technical reasons, including but not limited to signal interference;"

Just about anything!

"subsequent changes in system design which precludes Licensee's operation on the applicable Premises;" *Sure. . . the guy across the street is giving them a better price.*

"Either party may terminate this License Agreement upon 90 day written notice with or without cause to the other party." *No way! You reserve the best part of your tower for the communication company, turning away two (2) other companies who wanted the same spot. They build a bigger tower near yours and want your client to jump on their facility.*

"If Licensee determines at any time that the Premises are not appropriate under Licensee's design or engineering specifications for its communications facility or the communications system to which the communications facility belongs..." *They are waiting for a better deal to say goodbye to you. No facility is 100% effective. The second part of that. . . must have some "hidden meaning".*

"If ("anycom") determines at any time that the property is not appropriate for locating" *Hey guys do that before you sign the licensing agreement.*

"Tenant has failed to satisfy any necessary authorization or radio engineering criteria to use the leased property as contemplated in this lease" *Failure of a tenant should not be a reason to break a lease, cross this one out!*

"or. . . for any other reason, in its sole discretion, determines that it will be unable to use the site for its intended purpose" *Why have a License Agreement at all?*

"upon 30 day notice for any reason...tenant determines in tenant's sole discretion that the Premises are unacceptable from a business, technological, environmental or hazardous. . ." *Wow! Now there is one you don't want to sign. Hazardous? A new wrinkle. Are they afraid that the tower is going to fall over on their equipment? Cross this one out!*

When you have a few clients you will actually know who the companies are who will try these clauses "on" you. Be assured we have revised every one of them to the more traditional protections. No one wants to stick it to a client with a truly unsuitable site. Hard feelings can cause all sorts of problems but I assure you the research which goes into your site before they select it for licensing is more than adequate to answer all these clauses as *outs*. In the unlikely event they truly want out of a deal, remember you have reserved a spot on your tower for this client for a time, to the exclusion of another client who would have rented the same spot. You are entitled to some settlement in the process. Ask for it. And now my favorite one. This one sounds so good except for one clause. Can you find it?

"Licensee may terminate any Site Agreement without further liability on prior written notice to Licensor which shall become effective thirty (30) days after the date notice is mailed as follows: (i) changes in local or state laws or regulations which adversely affect Licensee's ability to operate on the applicable Premises; (ii) FCC ruling or regulation which is beyond the control of the Licensee and which renders the Premises not suitable; (iii) technical reason, including but not limited to signal interference; (iv) subsequent changes in system design which precludes Licensee's operation on the applicable Premises; (v) Licensee's inability to obtain any required license permit or approval which may be required for the construction and operation of Licensee's installation on the applicable Premises, including where the inability is caused by engineering surveys or structural reports; and (vi) for any economic reason."

If you got the answer (vi) to this one you are now an expert in Antenna Site Administration.

7.7 A Sample License Agreement

It is not my intention to pick out every clause in this sample License Agreement and *beat it to death* as was done with the termination clause. However the exercise does give you the benefit of seeing how different companies deal with the same issues. If you use common sense you will develop a skeptical and protective sense towards any problem areas in the communication company proffered agreement. If **not** using their agreement causes too much strain during negotiations, it really isn't that difficult to modify and sign off on the clauses which are of concern. The idea is to watch for the *hooks*.

If you reach an impasse, put a price on it. For example. While negotiating a license agreement, the corporate attorney for the communication company was adamant that the annual escalation clause of 5% was "outrageous" and that only 2% per annum would be authorized. "Oh" . . .says me; " I'll accept 2%. . . but then we will make the term of the agreement three (3) years". So now the ball is back in the communication company's lap. Do they want to lock up a site for twenty-five (25) years and know precisely what it will cost? Or would they rather renegotiate the lease every three years and have no long term guarantees on the price? The purpose of the escalation clause is to *keep pace with the marketability* of the site, inflation, and the ever closing window of site building opportunity. There is also the fact that after a few years the site might be billing a million dollars a month on behalf of the communication company. Five percent (5%) is not much of an increase with the income being generated by the (your) location. It also allows us to start out with a lower initial monthly rent until profitability is reached. Do not sign a long term license agreement without an escalation clause of some sort. Here is a sample License Agreement:

ANTENNA SITE LICENSE AGREEMENT

This (roof-top/tower) Antenna Site License Agreement (the Agreement) is made this _____day of______, 20___, by and between (Site Owner) having an address of________________(Licensor) and (Communication Company) having an address of ________________ (Licensee).

1. Premises. Licensor hereby leases to Licensee and Licensee hereby leases from Licensor the right to attach its antennas and microwave dishes (or satellite dishes) to the(roof-top/tower) located __________________,
City, County, State, Zip Code,
Latitude: ___D___M___S
Longitude: ___D __M___S,
together with approximately ______square feet of space located (within/or on the grounds either in your building or for their shelter). The space on the (roof-top/tower) together with the storage space is hereinafter referred to as the premises (the Premises). The rights conveyed hereby include the right to attach, operate, maintain and replace (equipment) on the (roof-top/tower); and, the right to install, store, operate and maintain related communications equipment in the (storage area or Licensee provided shelter) or other such structures as Licensee determines are necessary (the Structures). All of the facilities referenced in this paragraph are described in detail and illustrated in

a sketch of proposed construction which is attached and made part of this Agreement designated as Exhibit A. The Premises are leased for the express purpose of constructing and operating a telecommunications facility by the Licensee.

2. Term. The term of this Agreement shall be for a period of five (5) years from (date) (the Commencement Date). Licensor grants to Licensee the right to renew this Agreement for four (4) additional terms of five (5) years each on the same terms and conditions as are herein stated except that rent on the same terms shall be adjusted as provided below. Should the Licensee not wish to exercise the automatic renewal option, Licensee shall deliver written notice of Licensee's intent **not to renew** this Agreement no less than 90 days prior to the expiration of this Agreement otherwise this Agreement **will automatically** renew for another five (5) year term. The initial term and all renewal terms are referred to herein as the "Term".

3. Rent. The rent for the first (number) year(s) of the Term shall be $________without setoff or deduction, plus applicable sales tax, per month, paid monthly in advance, which Licensee shall pay to Licensor at such place as Licensor shall designate to Licensee in writing. If the Term does not begin on the first day or end on the last day of a month, the rent for that partial month shall be prorated by multiplying the monthly rent amount by a fraction, the numerator of which is the number of days of the *partial month included* in the Term and the denominator of which is the total number of days in the full calendar month.

(a) Beginning with the ______(nd/rd/th) year of the first Term and every _______(nd/rd/th) year thereafter, the then current monthly rental fee shall be adjusted by multiplying it by the the change in the Consumer Price Index (CPI) for the immediately preceding year for which the rent has remained constant. In no event shall the monthly increase for any adjustment period exceed ____% of the monthly rent established for the previous

Rent Term.

For purposes of calculation, the CPI used could be further defined as the Consumer Price Index-U.S. City Averages for Urban Wage Earners and Clerical Workers (1982-84=100) published by the United States Department of Labor, Bureau of Labor Statistics. Actually there are several CPI "up ticks" which can be chosen as the basis for escalation. Be careful to select one which reflects your best interest. Another approach would be as follows:

(b) Beginning upon the Commencement Date, Licensee shall pay to Licensor for this Agreement a monthly fee of $_______without setoff or deduction, plus applicable sales tax, during the first year of this Agreement. Thereafter the monthly fees shall be increased on each anniversary of the Commencement Date by the **greater** of five percent (5%); **or** the increase in the Consumer Price Index for All Urban Consumers.

4. Site Access. Licensor hereby grants to Licensee the right for ingress and egress, *(not regress!)* for utilities (connecting wires) necessary for Licensee to place its equipment in operation on the Premises. It is understood by the Parties that this should be by the most direct route to insure good engineering design.

Licensor further grants to Licensee the right-of-way for ingress and egress over across and through the (roof-top/building) or (land/shelter/tower) for the construction, operation and maintenance of the Structures on the Premises; and for the installation, construction and maintenance of telephone and power lines in connection with its use of the Premises; and for full access to the Premises from roads public or private. The term for this right-of-way shall commence upon execution of this Agreement and shall continue until the last to occur of (i) expiration of the Agreement Term, or (ii) removal by

Licensee of all of its property from the Premises after expiration of the Agreement Term which in no case will be longer than five (5) days from the expiration of the Agreement Term.

The location and configuration (rules) of the Right-of-way shall be agreed upon by the parties not later than ten (10) business days after the date of execution of this Agreement. This Right-of-way shall be included in any recorded Memorandum of this Agreement.

5. <u>Quiet Enjoyment and Title.</u> Licensor represents and covenants that Licensor owns the Premises and property subject to the Agreement in fee simple, free and clear of all liens, encumbrances and restrictions of every kind and nature, except for those which currently appear in the chain of title and are reported as exceptions on the commitment for title insurance which the Licensee may obtain.

   Licensor represents and warrants to Licensee that Licensor has the full right to make this Agreement and the Licensee shall have quiet and peaceful possession of the Premises and right-of-way throughout the Agreement Term.

6. <u>Subordination.</u> This Agreement shall be subject and subordinate to any and all mortgages, deeds of trust and other instruments in the nature of a mortgage now or at any time hereinafter a lien or liens on the property of which the Premises is a part, and Licensee shall, when requested promptly execute and deliver such written instruments as shall be necessary to evidence such subordination.

7. <u>Government Approvals and Compliance.</u> During the Term of this Agreement, Licensee shall comply with all applicable laws affecting the Premises, the breach of which might result in any penalty to Licensor or forfeiture of Licensor's title to the Premises. Licensee shall obtain any necessary governmental licenses, permits or authorizations required for the construction and operation of the Structures on the Premises and shall comply with

government regulations applicable to its operations including those of the FAA and the FCC.

Non-Ionizing Radiation Rules & Regulations. Licensee shall at all times comply with any federal, state or local law, rule or regulation pertaining to non-ionizing radiation, currently enacted or laws rules or regulations that are enacted, changed or promulgated. Compliance shall be on an individual basis and as it pertains to the entire facility. It is the Licensee's responsibility to provide documentation to any agency requiring such documentation for permission to operate any RF generating equipment and to prove compliance with any law, rule or regulation. Should any law, rule or regulation change, Licensee must, at its sole expense, take any and all steps necessary to comply with said change as required including, but not limited to a reduction in effective radiated power lobes, change in antenna type or change in radiation patterns necessary for compliance.

8. Assignments and Subleasing. Licensee may not sublet the Premises in whole, in part or within its system operation without Licensor's consent, and the making of such sublease shall not release Licensee from any of the Licensee's obligations hereunder. Licensee shall not assign or transfer this Agreement, or any interest herein, without the prior written consent of Licensor which shall not be unreasonably withheld *(be sure to delete "delayed or conditioned" if they try to insert it)*, and a consent to an assignment shall not be deemed to be a consent to any subsequent assignment. Licensee is expressly permitted to assign its rights and responsibilities under this Agreement without obtaining Licensor's consent to (communication company name), its successors in interest, assigns or any affiliate thereof. *(A new twist to this right must be carefully observed. Some communication companies are capable of running two systems from the same box and antenna. What they do is rent your location, install their box and antenna then provide transmission service to another communication company. If you don't pay attention your system will be serving several communication companies under the umbrella of one*

communication company. Exhibit A tells the story . . .but be sure to ask the question. If more than one system is capable of operating at your site it is certainly ok if it is within the system operated by the communication company you are Licensing. However, sub-licensing on the part of your Licensee should not deprive you of business. One more thing. If it does happen that this is taking place, it usually is for a "turn-on" or "start up" period. After a time the system installed usually will have to be enhanced (more boxes) or split out. Be ready for this eventuality by making attractive terms in your License Agreement to both entities.

9. <u>Notices.</u> All notices, demands, requests, consents approvals and other instruments required or permitted to be given pursuant to this Agreement shall be in writing, signed by the party and shall be deemed to have been effective upon delivery if served personally, including but not limited to delivery by messenger, overnight courier service or by overnight express mail, or upon posting if sent by registered or certified mail, postage prepaid, return receipt requested and addressed as follows:

To Licensor: (Name & Address)

To Licensee: (Name & Address)

Sometimes: (with a copy to Name & Address)

The address to which any notice, demand, or other writing may be delivered to any party as above provided may be changed by written notice given by such party as above provided.

10. <u>Licensee Improvements.</u> Licensee shall have the right, at its sole expense, to make such improvements to the Premises as it may deem necessary for the construction, maintenance and operation of a telecommunications facility site. All licensee's improvements *on* the premises including but not limited to all Structures, shall remain the property of Licensee. Except that nothing in this paragraph is to be construed as to permit the Licensee to

expand its facilities beyond what is herewith defined in Exhibit A, attached hereto, without the approval of the Licensor. Upon termination of this Agreement, Licensee shall, to the extent reasonable, restore the Premises to its condition at the commencement of this Agreement except for ordinary wear and tear and damages by the elements or damages over which Licensee had no control. Licensee and Licensor agree that Licensee shall not be required to remove any improvements which are permanently affixed to the land or Structure, including but not limited to foundations, footings, concrete paving, gravel, vegetation or utility installations which have not be rendered unusable by the Licensee or its agents when removing its property. *The communication company builds a 100' microwave tower then leaves your property. They remove the tower and "cut the mounting bolts on the foundation". Now they have to remove the foundation because you won't be able to use it.*

11. Interference. Licensee shall be responsible for coordinating and resolving any interference problems with exiting transmitters or receivers on or near the (roof-top/tower) within one month of the commencement of this Agreement; and any transmitters or receivers that may, through a revision of this Agreement, be added at any future date. Licensee shall not be responsible for coordinating or resolving any interference problems that may result from Licensor's subsequent licenses of attachment or space leased to another other party *after* the Commencement date of this Agreement. Licensor and Licensee both agree to notify the other party whenever different or additional transmitters or receivers are planned to be added to the (roof-top/tower). *Licensor will apply this regulation uniformly to all Licensees co-locating at the Premises to insure an effective and amicable resolution of all interference problems.*

12. Operating Utilities. Unless otherwise defined in Exhibit A, Licensee shall fully and promptly pay for all electricity, heat, light, gas, telephone service or other utilities furnished to the Premises and used by Licensee throughout the term hereof, and for all other costs and expenses of

every kind whatsoever in connection with its use, operation and maintenance of the Premise and all activities conducted thereon. Licensee shall supply its own back-up power source *unless otherwise designated in Exhibit A.*

13. Maintenance. Licensee shall maintain the Premises in good condition and state of repair. Licensor shall maintain its property in good condition and state of repair to avoid interference with Licensee's use of the Premises. *Weed and trash control!*

14. Hold Harmless. In no event shall Licensor or any of its partners, officers, agents or employees be liable for any injury to persons or damage to or theft of any property unless and to the extent such injury, damage or loss is caused by the gross negligent or willful misconduct of Licensor. Licensor shall not be liable for any such damage caused by other Licensees or Sub-licensees or persons in, upon or about the Site, including any of the Licensed Premises, or caused by construction of any private, public or quasi-public work, or any incidental, special punitive or consequential damages arising out of any loss of use of the Licensee's system or any part thereof, even if due to the gross negligence or willful misconduct of Licensor. In addition, Licensor makes no representation or warrantee as to the suitability of the Site including the Licenced Premises for the permitted use hereunder. *We do not want to be liable!*

15. Insurance. Licensee shall, at Licensee's own expense, maintain throughout the Term of this Agreement combined single limit policies of bodily injury and property damage insurance with a limit of not less than $1,000,000 per person, $1,000,000 per occurrence and $1,000,000 for property damage insuring all liability arising out of such party's use, occupancy or maintenance of the Premises. Licensor shall be named as an additional insured on Licensee's liability policy. Licensee shall have on file with the Licensor a current certificate or other evidence of insurance coverage during the Term of this Agreement. In addition, any contractor engaged by Licensee to perform any work on the Premises on behalf of the Licensee must

carry similar insurance and comply with the provisions of both Paragraph 14 & 15 while such work is in progress.

16. Termination. *See Section 7.7 of this Chapter.*

17. Binding on Successors. The covenants and conditions contained herein shall apply to and be binding upon the heirs, successors, executors, administrators and assigns of the Parties hereto.

18. Access to Premises. Licensee, its employees, agents and contractors shall have full access to the Premises during the Agreement Term, consistent with the Licensor's standard Site Rules And Regulations for Access and Security attached and made part hereof as Exhibit B.

19. Governing Law. The Parties intend that this Agreement and its relationship of the Parties will be governed by the laws of the (State/Commonwealth).

20. Entire Agreement. All of the representations and obligations of the Parties are contained herein. No modification, waiver or amendment to any conditions or provisions of this Agreement shall be binding upon any party unless made in writing and agreed to, by the Parties or by the duly authorized agent of the Parties empowered by a written authority to sign for that Party. The waiver of any party of a breach of any provision of this Agreement shall not operate nor be construed as a waiver of any subsequent breach of that provision by the same party, or of any other provision or condition of the Agreement.

21. Headings. The headings of paragraphs and subsections of this Agreement are for convenience and reference only and shall not be deemed to limit, construe, affect nor the meaning of any sections or subsections.

22. Time is of the Essence. Time is of the essence for Licensor's and Licensee's obligations under this Agreement.

23. Severability. If any section, subsection, expression or

provision of this Agreement or application thereof to any party or circumstance shall, to any extent, be invalid or unenforceable, the remainder of said section, subsection, expression or provision of this Agreement, or the application of same, to parties or circumstances other than those to which it was held invalid or unenforceable, <u>shall not be affected</u> thereby and each remaining section, subsection, expression or provision of this License Agreement shall be valid and enforceable to the fullest extent permitted by law.

24. <u>Broker Relationships.</u> Except for:

(disclose any agency here)

Licensor represents and warrants that Licensor has not signed a listing agreement, dealt with or otherwise agreed to pay any commission, finder's fee or other like compensation to anyone in connection with this License Agreement without full agency disclosure to Licensee. Licensor also agrees to indemnify and hold Licensee harmless from any fee and against any such claims or costs, including attorney's fees, incurred as a result of the transaction contemplated by this Agreement.

25. <u>Additional Assurances.</u> Each of the Parties agree to do such further acts and to execute and deliver such additional agreements and instruments as the other may reasonably require, to consummate, evidence or confirm this Agreement or any other agreement contained herein in the manner contemplated hereby.

26. <u>Right to Record.</u> Upon the request of the Licensee, Licensor agrees to promptly execute and deliver to the Licensee an acceptable <u>Memorandum of License</u> in recordable or registerable form setting forth the *general terms* of the Agreement and such other information as the Licensee shall request to establish its entitlement to this Agreement. This memorandum will not place into the public record any information deemed of a competitive or propriety nature to the detriment of the Licensor's business operations.

27. <u>Interpretation.</u> The Parties to this Agreement (and

their respective legal counsel) have reviewed and/or have revised this Agreement to the satisfaction of the other party.

28. Date of Agreement. The Parties acknowledge that certain obligations of Licensee and Licensor are to be performed within certain specified periods of time which are determined by reference to the date of execution of this Agreement. The parties therefore agree that whenever the term "date of execution of this Agreement" or words of a similar effect are used herein, they shall mean the date upon which the Agreement has been fully executed by the Licensor and Licensee whichever is the later to so execute this Agreement. The Parties agree to specify the date on which they execute this Agreement beneath their respective signatures in the space provided and warrant and represent to the other that such a date is in fact the actual date on which each duly executed this Agreement.

IN WITNESS WHEREOF, The Parties hereto have executed this Agreement on the day and year first above written.

LICENSOR:

____

By:
Title:
Date:

Notary:
I,_______________do hereby certify that:

____

personally appeared before me this day and acknowledged the due execution of the foregoing instrument.

Witness my hand and seal this ___day of_____, 20___.

____

LICENSEE:

(same format)

The above sample agreement might look familiar to many in the industry who review it. It should. It consists of bits, pieces, fragments and phrases from countless agreements on file both in effect and expired which I have used at my own site. It also shows many of the countless revisions and "haggles" over terms and conditions negotiated over the years. In general, the format is patterned from recent communication company agreements with additions, comments and insertions added, when seemingly in my humble opinion were deemed appropriate (because I got stung once that's why)! Also, I wish to express my appreciation to the many legal minds past and present whose thoughts I have embraced. They continue to contribute so much to the wording that confounds some and illuminates others. So kudos to just about everybody who every put their "two-bits" into one of these specialized agreements. The proof is in the execution thereof.

The agreement sample is but a starting point and is not intended as an actual agreement to be used with impunity. In order to create a workable licensing agreement from scratch or as a point of reference in the review of a proposed draft on the table, you will find that it is distinctly removed from the more traditional lease agreements found in transaction guides which are likely to surface if your attorney has had little experience with communication issues. As always, seek expert advice when reviewing someone else's agreement or devising your own. And above all, **do not be reluctant to ask questions** or to request clarifications. Often these questions alone provide revised text proposals without confrontation. Good luck.

7.8 Antenna Site Rules and Regulations

The final section in this chapter presents antenna site rules & regulations as applied to access, security issues and procedures for all users. In the Section 7.7 above, the agreement draft designated these rules in an exhibit mentioned in item 18. These are *your* rules. And should be applied unilaterally *to all* your tenants. They may vary between sites particularly if it is a roof-top or tower location. But essentially they are common sense items which establish the justification for uniform management and application of responsibility. In apartment house leases items such as the playing of radios too loudly, running noisy equipment or control of pests is of concern . Similar items apply to antenna sites and should be included in the Antenna Site License Agreement as a matter of your dedication to a good and proper administration of your site.

Here is a list of items taken from a number of site rule exhibits which should help you make up your own exhibit. Even if your site agreement was provided by the communication company you still have the right to insert your own rules as an exhibit and to insist upon compliance.

1. Doorways, vestibules, halls stairways and other areas in and around the building shall not be used for the disposal of trash or be obstructed by Licensee or used by Licensee for any other purpose than entrance to and exit from the equipment area.

2. Licensee advertisements will not be permitted in the equipment room or on the outside shelter or building walls. Property signs are permitted. Each cabinet in the common area shall be identified with a professional card indicating the engineer of record, duty engineer or who should be called in the event of an emergency or problem. FCC licenses shall be posted and current in those services requiring this posting.

3. Licensees will make no alterations or physical changes in or to the equipment room without the written permission of the Licensor.

4. When entering or leaving the equipment room with any large piece of test equipment or other large items, the freight elevator will be used. If the public elevator must be used it will be used only when regular building tenants are not using it. Licensor will determine the method and moving of such large items so as to ensure the safety of all concerned and that potential harm to the building be minimized. Advance notice of at least forty-eight (48) hours is required for the routine movement of any large equipment. Emergency moves will be dealt with on a case by case basis.

5. Licensor shall have the authority to prescribe the maximum weight imposed upon any area and the manner in which equipment is placed.

6. Doors to the equipment room or mechanical rooms shall not be propped open due to air flow/air conditioning and fire regulations.

7. Use of keys or access cards is limited to use by Licensee. Any Licensee, "loaning keys or cards" to other Licensee's or other individuals <u>not so authorized with access authority</u> is in violation of the License Agreement. Loss of the card or keys issued to the Licensee will result in charges for the replacements.

8. No animals, pets, birds, children or bicycles are allowed in the equipment room. In addition all food, soda or other food stuffs and wrappers shall be removed from the site. Firearms are not permitted at building locations. Firearms carried for protection are authorized at remote sites if permitted by the local municipality and registered in advance with the Licensor.

9. Licensee shall not tamper with or attempt to adjust any temperature control thermostats in the equipment room. Licensor shall adjust thermostats as required to maintain equipment room standard temperatures.

10. Licensee will comply with all requirements necessary for security of the equipment room and the antenna site during business hours, after hours and on weekends and holidays. Any intruder investigation resulting from failure to de-activate an alarm will be automatically invoiced to the Licensee.

11. Access to the equipment room is limited to individuals having need to perform work in that location. Licensee's wishing to escort guests to the site shall schedule such visits with Licensor in advance of such visit which will be during normal business hours.

12. All routine service calls will be scheduled between 8 AM and 5 PM weekdays.

13. Licensor reserves the right to rescind any of these rules and regulations and to make other such rules and regulations, if in its judgement, from time to time, they are required for the safety, protection, care and cleanliness of the antenna site/equipment room, the operation thereof, the preservation of good order therein, and the protection and comfort of the Licensees, their agents and guests. Such rules and regulations, when made and with written notice thereof given to the Licensee, shall be binding upon Licensee in a manner as if originally herein prescribed. Licensee shall cooperate fully with Licensor in regards to all of the above, as well as to future rules and regulations pertaining to the equipment room, antenna site and general building premises.

14. Licensee shall also adhere to standard building rules and regulations.

CHAPTER EIGHT

8.1 The Zoning Process: *No Fight, No Site!*

In the past several years over $20 billion dollars has been paid into the U.S. Treasury by the stockholders of telecommunication companies spurred on in the hopes of a profitable return for competitive wireless telephone services. The investment rationale for all this excitement is that once the money is raised and the construction completed, high financial returns will result.

At the time the new PCS service was initiated, a sweeping new law in the form of the *Telecommunications Act of 1996* was also passed. Antenna siting was made a part of this new law by attempting to prohibit any action that would discriminate between different providers of personal wireless services. It also was supposed to prohibit any action that would ban altogether the construction, modification or placement of towers or antennas in a particular area.

The antenna siting section, *Section 704,* seemingly followed the same logic that the FCC used over the past two decades. Both the FCC and most state regulatory commissions previously exercised jurisdiction over the rates and coverages of the mobile telephone (and paging) industry. Because it was an offshoot of the telephone system and interconnected to the *public telephone switching network,* PTSN, the services were defined as *public utilities.* Federal Regulations applied to interstate activity while local regulation concerned itself with the nature of intrastate matters. *Certificates of Public Convenience and Necessity* were issued and tariffs published and controlled for every communication service offering from dispatch, paging and telephone interconnection *(patch)* service. When communication companies were public utilities, they also had the unique advantage of having limited power of *eminent domain.* The effect of this public utility right was that it was only necessary to prove that the construction of the tower was in the public interest. Although rarely used, *radio common carriers,* RCC's, at one time actually had the right to request condemnation of real estate or to obtain easements in order to construct towers just about anywhere they were needed. This authority still exists today to some extent for both telephone and electric power companies many of whom still operate under public utility laws. However, when radio common carriers were *deregulated,* they lost their power to put up towers under any notion of eminent domain. It wasn't much of an issue in the past because most radio towers were erected on high hills and there was no need for so many of them. But the right of eminent domain has been used for years by the power companies to put towers, poles and sub-stations literally anywhere they are needed.

Compare the communication carrier of today needing radio towers with the power company. There are many similarities, i.e. pole placement, attachments, potential for radiation, chemical or electrical hazzards, aesthetics, ground space for sub-station and distribution equipment, security, access and public acceptance to name a few. Now compare the *number* of poles/towers allocated to each. What could this ratio be? One radio tower per a thousand power poles? Even if the ratio is totally inaccurate the question remains, why do we have so many regulations directed to so minor a problem. The question in its broader sense may never be answered. However if you want to build a tower you are going to have to get a *building permit* and to do this it might be necessary to secure *zoning approval* at the location you have selected.

In the early 70's more and more communication companies started to receive FCC licenses for services whose operation was similar to that of the telephone company. An unregulated business band frequency could provide the same kind of mobile telephone or paging service that a rate regulated carrier could provide. It was soon realized that any frequency, albeit business, government or even commercial users could provide like services to the *public*. Regulators then shifted their attention to the definition of the word *public*. For a time the utilities could serve the *general public* but the business band carriers were only permitted to solicit a *specific public*. Well try enforcing that. People buy the services they want. With so many suppliers of similar telecommunication services in a given area why should any one be a monopoly to the exclusion of the others?

Nevertheless, in an attempt to hold on to their monopoly position telephone and independently regulated radio companies spent years in *competitive hearings* attempting to enjoin unregulated operations. Millions in legal fees were spent but eventually the dust settled. A monopoly can not exist when so many competing firms in the same market provided essentially the same services. The legal arguments to retain monopolies (as well as to open up the markets to free competition) were in many cases brilliant. They may someday be memorialized in volumes of legal texts. But no one much cares any more because nothing applies. *Deregulation* was applied to the so called radio/miscellaneous common carrier when the FCC created the cellular service in 1982 with the release of the cellular

FCC Docket. One of the results of this creation categorically *preempted* states from regulating any of the new industry at all. Other exemptions soon followed. State and local governments were literally pushed out of the radio common carrier business all together. But the local regulatory bureaucracy did not die as the need for thousands of new radio towers has apparently awakened interest once again.

It was reasoned that the construction of anything within a municipality by a business entity (as opposed to a public utility) can be covered by local building codes, zoning and land use ordinances. If none are in place they are enacted. If any new regulation is in conflict with Federal Law, it is merely redirected to apply to the interpretation of local concern. If an existing tower built in prior years happens to sit in a newly defined restricted area (zoned) it is *grand fathered* sometimes with regulations that prevent tower augmentation or any necessary improvements. This seemingly self validating segment of our governing process has always been based upon the common good. The process of regulation and enforcement can however, sometimes take some interesting twists. A slow down in governmental processing can decimate a comprehensive communication plan if a single but mutually exclusive component of a system is frustrated or can not be constructed. By talking the issue to death, inviting clarification after clarification, and moving the agenda from month to month, the approval to construct can literally die without anyone making a decision. It is not so bad if the process involves a simple build-out or a cell splitting strategy. But the process can be devastating if it is the *initial* system design that is being stymied. In a hostile zoning environment an application may capture some headlines for a time, be delayed for a time, and then be withdrawn after a time. However some zoning boards are very efficient, hospitable and even sympathetic to the new industry and actually direct fast approval and complaint disposition through effective rules of practice. It is important to know what kind of zoning board you will be up against.

Indirect and specific requirements enacted locally make the job of opposing or exemption by Federal regulation even more difficult. Sometimes referred to as a moving target, a local zoning board may focus upon the *look* of the facility and not its function. Enacting an ordinance that may call for referring the proposed tower to a voluntary and artistically *qualified* concerned citizen group to hear their recommendations about its proposed presence in the community. This orientation has resulted in some pretty funny compromises. Towers painted blue to "blend into the sky" or green up to the tree line and blue after that. There are many other restrictions which once installed give testimony to this process all over the country.

There are criticisms that in some instances a zoning process may not be fairly administered, even going so far as being prone to influence peddling, intimidation, steering or kickbacks in the form of telecommunication service *tradeouts.* In one case the process has been referred to as a "not so benevolent dictatorship". With so many rulings having been made through the years it would be unrealistic to hope for the extinguishment for these regulations as being unconstitutional. The matters before zoning boards are typically *issue* driven. Keep this mind and adjust your attitude accordingly.

The FCC was quick to preserve local zoning authority. But they also clarified a position that local zoning authority *may be preempted by the FCC*. Zoning delays do not violate this part of the law. The area is cloudy at best with the effect of creating *market entry barriers* anyway. The introduction of hefty franchise, processing or permitting fees, although not illegal, are discouraging when compared to other similar but non-radio related applications. A permit in one municipality to construct a half million dollar building might only be $500, with the same municipality charging a $3000 non-refundable permit fee for a shelter. These effects give the appearance of discriminatory practices and in some instances are seemingly intended that way. Could it be that with all the billions of dollars going to the Federal Government, the localities are resentful that they are not in for a *piece of the action* as some have speculated? The concerns and decrees forming the basis for the power from which zoning boards derive their authority can be dealt with systematically until either a project is approved or all the grounds for rejection have been exhausted in the courts. In general most communication companies have not opted to challenge nor enrage but have been as accommodating as possible, realizing that every case *won* is precedent and every case *lost* can or at least hopefully fall back to an alternative plan. But from time to time these Titans they do clash. And it makes for interesting headlines.

In retrospect, it is amazing how effective the *grapevine* seems to be with zoning boards everywhere. Proposed ordinance text *pops* up all over the country seemingly instantaneously, complete with case records and supportive discussions of what areas to emphasize in order to maximize local TV coverage. Some ordinances have been worked through locally and are fairly comprehensive, others order up what seems to work elsewhere, rubber stamping the process with a few modifications here and there. And some of the modifications are real wingers. We have mentioned a few of them in past chapters, fall zones, aesthetics, unproven speculations of the effects of non-ionizing radiation on people. Some zoning board members who in the past worked with residential and commercial developers without much public attention now seem to be caught up in a greater enchantment for higher office from all the publicity resulting from protecting a community from radio towers.

It has been pretty much accepted that Section 704 of the Communications Act of 1996 as it was written has proven impotent. *An industry assumed, by Congress to be able to proffer billions of dollars for the sale of spectrum has been brought to its knees*, as Jeffery Silva, a writer for *RCR Focus* put it, *by a potent, eclectic mix of soccer moms, zoning czars, environmentalists and union members*. In some cases the communication company, entering an application to construct a simple monopole in some out of the way back lot of a manufacturing plant, is made out to be a *killer of infants*. Splash and suggestive headlines follow weeks of attempts by the communication specialist to satisfy every request made by the zoning board members in seeking final approval for the project with no guarantee of success.

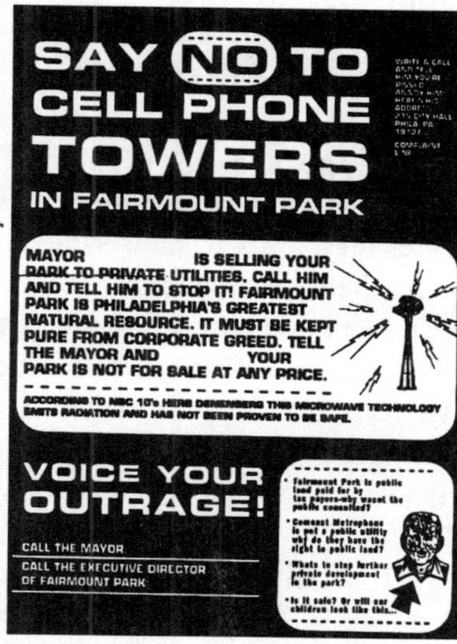

FIG. 8.1 FLYER IN OPPOSITION TO A TOWER APPLICATION

I once found myself at a local college extension taking a graduate public administration course in community planning from a former zoning officer. His insight into the *logic* justifying some of the many monumental zoning decisions while he served on the board were simply astounding. This course probably was the reason I abandoned any further enthusiasm for obtaining a Masters in Governmental Administration.

Which brings us to the issue of should you establish your own case before a zoning board or use a hired gun? In my opinion it doesn't much matter. The process will exhaust you or your legal representative's clock time with little difference in the outcome. In addition, you might wind up with some appeal room if you lose on some technicality which later can be revisited to the exclusion of all the other matters already resolved. Give this some thought. I have found that zoning folks generally don't mind talking to *locals* in a manner not reserved for the formality of your legal representative. In general it has been my experience that they also seem a bit more tolerant of good faith efforts rather than volumes of legal discourse. Finally, another strictly personal observation is that there frequently seems to be a *good guy-bad guy* situation going on in zoning board meetings. This is where one board member is violently opposed to everything you want to do while another tries to protect your rights as a businessman and citizen who should be able to proceed with their plan. There is no business like show business.

8.2 Zoning Classifications

The antenna site zoning process has created a whole new industry of specialists. Often former zoning board members return as advocates to represent clients in their same area. What works may be different from municipality to municipality. As a resident of your community, tax payer, local businessman and antenna site developer it may be that you'll have an edge over the communication attorney who comes in ready to overwhelm everyone with legal prowess. All bets appear to be off at a zoning hearing probably because they make up their own rules and spend most of their time defining what is acceptable, adjusting the rules to fit the situation.

Zoning boards define *sectors* of their jurisdictions for various purposes. Commercial, residential, manufacturing, agricultural and recreational to name a few of the broader classifications. When it becomes necessary, subcategories are defined such as a distinction between light or heavy manufacturing or multi-family residential etc. In addition to accommodating years of confusion, it is possible to have multiple classifications such as a commercial and/or residential R-4, all defining the same property. Some areas have been targeted for reclassification with *use it or lose it* ordinances, one application which is forfeited to another due to inactive use. As an example. Back in the 40's a retired couple may have taken in a roomer to help make ends meet financially. The boarder's room may have evolved to a small apartment thus moving an R1 residential to an R2 residential. No one really cared too much about those situations in the past but now if the apartment *does not* meet certain codes (imposed by the same municipality) such as separate sewage lines, electrical service, bathroom standards, kitchen appliances etc. the R-2 residential status can revert back to R-1 and never again be used for a small apartment. An over simplification for sure, but the general idea is to discover what is important to your zoning board. You can learn much about these issues by visiting your board's document section usually available for public inspection during business hours. Ask a lot of questions.

The next level in the broad zoning classifications mentioned above are the lists of *permitted uses*. A typical *permitted use* regulation looks like this:

In a C-1 General Commercial District, no building or land shall be used, and no building shall be erected which is arraigned, intended or designed to be used for other than one or more of the following uses:

Then the *SIC Codes* are used:

411	Local & Suburban passenger transportation
43	U.S. Postal Service
4941	Water Supply
5993	Cigar Store
752	Automobile Parking
7832	Motion pictures, theaters, except drive-in
803	Offices of osteopathic physicians

These are just a few of the hundreds of *Standard Industrial Classification* codes (SIC) found in the current edition of a manual published by the Federal Government and available from the U.S. Government Printing Office, defining the activities permissible in this particular zone. Nothing is uniform from municipality to municipality however. Radio towers and antennas will rarely be on this list because as a *permitted use* there would be no need for a hearing. This does not mean to say you can not put a tower in this zone. It means because it has not been defined, a *variance* will now be necessary to proceed. So actually you may not be asking for permission to put up a tower at all but asking for a *variance* of an existing *permitted use* to include a tower in a given location.

Other classifications might be added as well. These might take the form of *conditional uses,* which in this case would be the same as a *permitted use* except that there are additional hoops through which you must go. Phrases such as "without first complying with appropriate provision (of other articles), including proper certification from the Board of Supervisors etc." is typical.

Here are some more examples of *conditional uses* to the previous

C-1 zone. See if you can predict why or what the "other applicable provisions" could possibly be:

552 Motor Vehicle dealers (used)
554 Gasoline service stations
557 Motorcycle dealers
5999 Miscellaneous retail stores not elsewhere classified (i.e. a dirty book store?)

But the zoning process is not finished with just *permitted* and *conditional use*. Oh no, not yet! Try this addendum to the above regulation:

Uses in this zoning district shall not be permitted on a property that abuts a zoning district which permits residential uses.

This phrase means that you may not be able to put a tower on a piece of land which might even be a *permitted use* for this purpose because it is adjacent to residential (R-1) property even if that property is three miles down the road.

How easy would it be to add a *radio tower* to the list of permitted uses? Well for one thing someone would be interested in the height of the tower, which obviously would vary, so it would automatically be relegated to a *conditional use*. Then again it might be subject to other regulations such as the aesthetic committee. In reality, it is unlikely that you will see it at all unless of course the municipality owns some land that they want you or your clients to rent from them.

There are many other sections contained in zoning ordinances which can give rise to concern even in areas which permit towers. These could be impossible to apply set back distances, proximity to other structures,

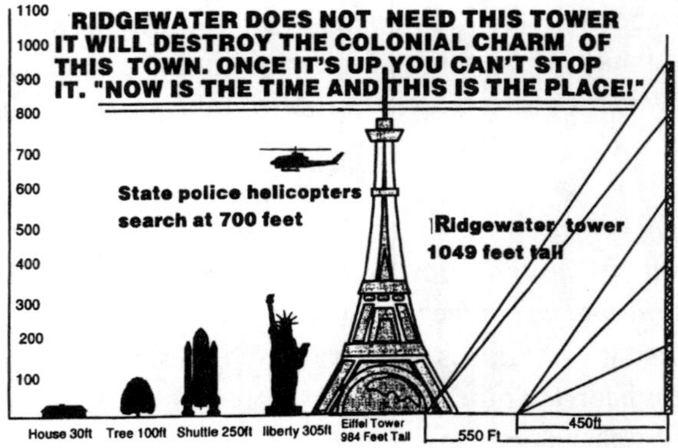

Fig 8.2 Organized campaign to oppose a tower based upon aesthetics.

height or material restrictions, sign or access road regulations, parking regulations, handicapped accommodations, landscaping, fences and service entrances. My favorite regulation was a requirement for a handicapped bathroom to be installed near the base of a mountain top tower.

8.3 Preparing the Zoning Application

The first step in preparing a zoning request is to get a copy of any ordinance effecting towers and the application form if they have one. It is also helpful to review a completed submission from another applicant who was successful in their bid to construct a facility. The following notes may also be helpful in preparing your submission which will more than likely be a request for a *variance* to an existing *permitted or conditional use* for the land upon which you plan to build.

1. As an applicant you must be fully prepared to explain your proposal, either in person, or by an authorized representative. You usually will be able to represent yourself and legal counsel may not be required.

2. Statements before zoning boards are usually made under oath, and every applicant or witness is subject to cross examination or general questioning by the board or any other interested party. If you don't know an answer to a question, say so. It is better to defer a question to a subsequent meeting than to fake an answer and get caught at the next meeting. Counter questions you do not intend to answer. *I understand Mr Tower Builder, that these antennas generate harmful radiation which cause cancer is this so?* (the TV lights come on) *Mr. Zoning Officer, what type of radiation are you referring to* (pause) *would that be ionizing or non ionizing? Well I don't know, I read an article in the Wall Street Journal* (it's a wonder he didn't say the Journal of American Medicine) *that these towers cause radiation,* (now looking directly into the cameras) . . .*and I believe it is in the interests of our citizens to know the truth about what you intend to build in our community. Well sir you are correct these towers do radiate a certain amount of radio frequency radiation and I would be pleased to give you a full report on any effects that this tower might have at your nearest convenience.* (TV camera lights go off). Move the issue to the next meeting by sending him a written report in a few days and oh yes, also send a copy to each of the reporters at the hearing. Next time it will be old news.

3. Zoning boards make their decisions based upon the *record* which is created by your presentation using all relevant facts. Be sure that all the good points are in writing illustrated by simple wording, lots of pictures, maps, some charts and a few graphs. Allow the not so good points to be developed through questioning and here say. These wind up on transcripts which few outside people look at if they are just there to scarf some publicity.

4. If a *variance* is approved, the applicant will receive a written decision from the solicitor of the zoning hearing board stating the terms and conditions of the approval. Any party aggrieved *I wish they would defined that but they seldom seem to* by a decision of the zoning board, may appeal such decision within 30 days as prescribed by local laws. Watch out, it could even be the *bad guy* on the board. As such don't expect a building permit which usually won't be issued until well after the appeal process date has expired.

5. *Variances* are <u>specific to your request</u> and may only be valid for a short period usually about six (6) months from the date that the thirty (30) day appeal period ends, unless specified or extended by the zoning board. If a building permit is not obtained within the required time, the approved *variance* application could expire. As such, to proceed with the proposal a completely new variance application might be required *of course*. Problems sometime arise from the fact that you have two (2) drop-dead dates. The building permit expiration and the *variance* expiration. Shipping steel can sometimes cause some problems in this area so planning is crucial to a successful project completion.

6. Sometimes a building permit can be obtained before the expiration of the thirty (30) day appeal by signing an affidavit stating that construction will cease if a complaint is received by the municipality.

A zoning application in many instances serves as a cover sheet for the various subsections of the proposal. One section might go to the engineer, another to the planning person, and yet another to the taxing authority. Prepare your subsections for easy separation and reassembly so that parallel processing is possible rather than one complete bound book which must be passed from department to department. Actually it is best to go visit each of the reviewing departments to see just what it is that makes them happy or bummed out about previous tower applications. This might sometimes be referred to as *ex parte* (gasp) so don't discuss *your* application, discuss a past application which they have previously reviewed (and liked). Then use that format. Here are some more thoughts:

1. When applicable, submit a sketch, a plot plan and a floor plan showing all existing and proposed property lines and structures with dimensions.

2. A cover letter explaining the nature of the project and the *variance* requested. State precisely why you cannot comply with the requirements set forth in the zoning ordinance and what hardship you feel will be overcome by the use of your property <u>within</u> the requirements of the zoning ordinance. *One favorite presentation is to show the need for three towers if you can't have the one you are applying for. "We need to build this big ugly tower here so we won't have to come back and build three others. . .here . . .here . . .and here!*

3. Usually if the application is incomplete or adjudged as incomplete it can be dismissed. Although not likely, it is always good to follow an outline and supply information which is expected. A simple *clear land title* description, for example is always required so don't forget the obvious. Here is a list of items which may be required in a typical zoning application or subsequent hearing:

   (a) Topographic Map
   (b) Zoning Map
   (c) Detailed Street Map
   (d) Aviation Sectional Map
   (e) Wetland/Conservation Map
   (f) Coastal Commission Map (if applicable)
   (g) Tax Assessor's Map
   (h) General Land Values in Market Area
   (i) Other documents applicable to the tower location
   (j) Document the following information:
      1. actual zoning (look for loop holes)
      2. high points
      3. towers, water tanks, other tall structures
      4. highways, streets and roads
      5. land owners (who own a lot of the surrounding land - look for "barons" seek their

approval)

6. rivers and streams
7. existing towers (why they can't be used)
8. airports and flight paths

(k) Make contacts with local building and zoning staff. Make contacts with other commissions, get concurrences including FAA or state aeronautical boards.
(l) Project market patterns and land values into the future.
(m) Confirm what approvals will be needed. Get as many as you can before the hearing.
(n) Obtain all forms.
(o) Coordinate hearing dates with building requirements.
(p) Review local opposition. Contact spokespersons and give them *all* your information. Head off any objections especially on hearing nights.
(q) Pay fees. Copy checks.
(r) Have specialists standing by for RFR, tower engineering or wetland questions.
(s) Know where the hearings will be held, check the room out, microphones podiums and times.
(t) Make a pleasant presentation, don't hide behind papers.
(u) Follow your outline and their format. Plug in reasons they can use to approve your request.
(v) If everyone looks happy with what has been said, shut up and sit down.
(w) Get the building permit as soon as possible.
(x) Dig the hole.

Many site acquisition companies have developed data bases from detailed check lists and field surveys which facilitate the zoning process in a given municipality. Since it needs to be done anyway the check list records many subjective bits of information in the event there is a personnel change during the process or if the staff has to go back later for another hearing. This data, if it could be collected and distributed throughout the country would be marketable and invaluable. It is obvious that you will need to identify subdivisions, county names, school districts and municipalities. In addition research the politics, the personalities, specific mailing addresses, e-mail addresses and telephone numbers (hopefully direct lines) of everyone and everything you will have to be dealing with. Titles vary all over the country so be sure to get the *exact* title correct. Try to obtain professional cards. Give them yours. In addition knowing your contact's personal assistant name and title can be very helpful when asked that question *Does Mr. Town Planner know what this is in reference too?* After a time, call screening won't be necessary if you are pleasant and don't take up too much of Mr. Town Planner's time.

Ask how long the entire process typically takes. Ask about restrictions, limitations, area ratios and fees. Get a list of zoning codes which apply to related and similar projects like oversized flag poles or water towers etc. Do they want dimensions in English or metric or both. Keep a notebook of all dates, calls, contacts and information, both received and sent. Let them tell you what you need to do. Take notes in front of them. Determine if it is a simple matter of a zoning *variance* or if land use processing is also required. Don't over prepare your exhibits. Keep things simple.

Use photo simulation if you can. This is the process of scanning several pictures of the actual proposed site from several different perspectives and distances; then *pasting in* an actual tower picture, printing the document then submitting it as an exhibit. Often the picture shown from a distance lessens the mental impact not too mention eliminating hundreds of pages of testimony.

8.4 The Building Permit

One can not begin to summerize all the requirements that may be necessary for obtaining a building permit once your zoning approval has been obtained. The reason is that every municipality has different rules and regulations, requirements and personalities. The zoning application contains generalized data (not too specific). But when applying for a building permit more specificity is required. The *wet* diagrams (signed by Professional Engineers) may be required along with your geo-technical and Phase I Study Report. Site surveys showing tower location, orientation, structure analysis, landscaping details, grounding, utility routing and host of other documents may be required. Certain key inspections will more than likely be necessary also and have to be scheduled. You dig the hole and connect the *rebar* and are now ready to pour concrete. You need the inspector to

Fig. 8.4 Rebar ready for inspection by code officer prior to pouring concrete

come look before your pour. You have a locally licensed electrician who runs the drops, panels and meter bases, another inspection. When the tower is erected you may have an inspector coming over to insure that *hard hats* are being used and equipment is operated properly. Remember, the inspector can stop your project at any time and request compliance to some regulation or another.

The permit will require payment of fees and have specific times attached for various phases of construction and inspection. Often it is possible to leave the permit at the municipality until it is required then "pull it", which starts (*civil start*) the completion date clock. Other procedures might include a *clearing permit* with the construction permit. This might involve what is done with the dirt, what trees need to be cut down and how they are disposed of and the actual routing of utilities or roads right down to the thickness and type of the crushed rock. All inspected! Inspection seems to be arbitrary in that after several projects, if you or your vendor/contractor have built up a good reputation, you'll not see an inspector for anything other than what is designated on the permit.

It is very important that you go over the permit in detail before a civil start. Any clerical error will stop your project immediately. Property descriptions could be in error. I've even seen a permit that permitted the construction of the base, the shelter, roads and electrical and telephone service, but failed to provide for the erection of the tower. In addition if your land is leased it is a good idea to go back and review your lease agreement for full compliance having your landlord sign off on the permit. Recall that usually the commencement date is the date construction begins so have a check ready for your landlord when you ask for his approval of the construction permit. Works every time.

Once you have your permit in hand ask your building inspector for a list of certified or licensed contractors and materials suppliers in your area. Often local people know the codes inspectors and can save you lots of time and money by insuring that things are done properly.

Copy the permit, post it at the job site and get to work. Final inspection will usually be a few weeks after you are on the air with your first client. Site debris and landscaping, signs and safety considerations should be obvious to your inspector. Take lots of pictures during the construction process. Should an inspection be missed on the base for example you'll have photographs showing the hole, *rebar* positioning and base materials. Without them you can prove nothing.

8.5 Working with a Siting Ordinance

Over the years I have collected quite a few complete siting ordinances for radio towers and transmitting antennas. Most of them start out pretty close to a standard format but here and there they depart widely from accepted standards and norms. In a municipality without a specific siting ordinance you can pretty much do what already has been done as past practice. However with a siting ordinance in place, you might find yourself at odds with a specific paragraph for what seemingly does not apply to your proposal. In other instances you might find that the municipality is exercising jurisdiction in an area over which they have no authority. An example. A township in Pennsylvania required that all new communication company antenna installations proposing to go on an existing tower, obtain a building permit. The permit application required information on the position of the tower and plot plan etc., It was essentially a re-submission of the actual documents that constructed the tower in the first place. In addition it required an RFR study of the *predicted* radiation effects of the newly added antenna in relation to the nearby area. Also an intermodulation study of the effects of adding the new frequency to the overall facility. And finally, the building permit would not be issued without an certificated (PE) antenna loading study to prove that the existing structure could support the antenna installation (engineering opinions were not acceptable).

Most unsettling was the fact that these requirements were directed to the tenant not to the tower owner. The building permit was required for the installation of additional facilities no matter how small as well as for new facilities. The ordinance had been enacted after public notice but without specifically informing the site owner, so it came as a big surprise to everybody. It also delayed the operational plans of a paging company in deploying its improved service quickly to the area. Here were the concerns. If you do not support your tenant's application he would be required to attest to things for which he really has no first hand knowledge. You wouldn't particularly want him to do that anyway. Finally, do you want to challenge the validity of the ordinance or just get on with it?

Here were the arguments along with the final determinations:

(a) <u>The RFR Study Requirement</u>

Proposed: Allow RFR readings to be taken by a calibrated RFR meter *prior* to the antenna installation (to establish the exiting RF floor level). Then take readings after the communication company installs its antenna and begins operating (the floor level plus the new antenna). The two reports would be submitted as a fully accurate exhibit proving compliance to this sub-paragraph. This procedure would have been highly accurate, quick, producing actual RFR levels and was inexpensive. The municipality however rejected the proposal. A manual study was ordered from a local college physics professor who calculated only a hypothetical RF level resulting from the use of the antenna. Upon closer inspection the final report may have excluded the transmission line loss and the null fill pattern characteristic of the actual antenna. The report by the professor cost several thousand dollars. Once installed the readings were taken with the meter anyway as originally proposed. The theoretical report wasn't even close. As it stands now the inaccurate report is the one that is in the record.

(b) Intermodulation Study.

Proposed: That the intermodulation study be "waived" as there was no justification nor need established by the municipality for such a study and that such matters were clearly a function of FCC regulations and *contractual commitments to cooperate among all tower tenants so co-located, should such intermodulation result in any evident interference.* Further, several federal communication clients at the site, were operating in frequency bands not under FCC jurisdiction, would not reveal specific channel frequencies and were even operating in a *spread spectrum* mode making any theoretical intermodulation prediction totally inaccurate. The waiver request was rejected and the study was performed using only those frequencies which could be identified. The study was worthless because less than 60% of the frequencies operating at the site were included in the calculations. To be accurate *all* the frequencies must be used.

(c) Tower Loading Study.

Proposed: That an engineering opinion be substituted for a full tower loading study. Rejected. A full tower loading study for this particular guyed tower cost over twelve thousand dollars. As recently as one year earlier, tower augmentations were made with modifications costing in excess of $75,000 to replace guy lines, enhance stress points, paint and condition the facility for ten (10) additional antennas. During this same period a very heavy television antenna was removed along with its corresponding transmission line. This antenna weighed well over 1700 lbs. The associated three (3") inch transmission line was estimated at over 1000 lbs. A logical argument should have been accepted that if you remove an existing antenna system weighing nearly 2700 lbs from the top of the tower you would certainly have sufficient loading to mount an 11 pound paging antenna in its place. In this case however the entire fall zone for the tower was on the tower owner's property which was fenced and fully insured for failure and casualty loss by all the tenants.

I think what is exasperating about this situation was that the township had written an ordinance they neither understood nor could logically rule upon. Had there been one knowledgeable person on staff, it is pretty obvious rational waivers could have been accepted. As incredible as it sounds, an informal comment was made *just submit what we ask for so it can be part of the record.* In actuality, the data on file hardly meets the ordinance but because there were no deficiencies in the inclusion list, the permit was granted in spite of the quality of the submission. Finally, making a tenant pay for and obtain a building permit for inclusion on an existing tower structure is a pretty scarey precedent demanding some sort of clarification. But who wants to take on a zoning board? Perhaps a solution to this situation would be an enveloping application to add ten (10) additional antennas so that when the time comes they can just be installed under a previously granted blanket permit.

8.6 A Typical Siting Ordinance

Should you be working under a recently enacted ordinance you will have to decide what is worth accepting and what is worth challenging. The chances are that if another antenna tower hearing has gone before you a lot of precedent will have been established that you can immediately lock on to and employ. Should you be the first one through the gauntlet you might have some challenges on your hand. Here is a typical boilerplate ordinance circulating the country *(with some comments)*:

BE IT ENACTED AND ORDAINED by the Board of Commissioners . . .this is hereby enacted as follows:

ARTICLE 1234

Telecommunications Signal Facilities

Purpose: In recognition of the quasi-public nature of telecommunications systems, the purpose of this Article is as follows:

(a) To accommodate the need for telecommunication antennas while regulating their location and number in the township.

(b) To minimize adverse visual effects of telecommunications antennas and antenna support structures through proper design, siting, and vegetative screening.

(c) To avoid potential damage to adjacent properties from antenna support structure failure and falling ice, through engineering and proper positioning of antenna support structures.

(d) To encourage joint use of any new antenna support structures to reduce the number of such structures needed in the future.

Definitions: The following words and phrases when used in this Ordinance shall have the meaning given to them in this Section unless the content clearly indicates otherwise:

(a) **Antenna Height** means the vertical distance measured from the base of the antenna support structure at grade to the highest point on the structure. If the support structure is on a sloped grade, then the average between the highest and lowest grades shall be used in calculating the antenna's height. *Good thing the tower isn't in Vail, CO.*

(b) **Antenna Support Structure** means any pole, telescoping mast, tower, tripod, or any other structure which supports a device used in the transmitting or receiving of a radio frequency energy.

(c) **Telecommunications** means the transmission between or among points specified by the user, of information of the user's choosing, without change in the form or content of the information sent and received. *Whew, where did they get this one?*

(d) **Telecommunication Signal Site** means a tract or parcel of land that contains a telecommunications antenna as the principal use, its support structure, accessory building (s), parking, and may include other uses and equipment associated with and ancillary to telecommunications signal transmission or processing.

<u>Use Regulations:</u>

A telecommunications site with antenna that is attached *oops don't you mean <u>will be</u> attached?* to a pre-existing communications tower, <u>smoke stack</u>, water tower, or any other structure *is permitted in all zoning districts*. The height of the antenna shall not exceed the height of the existing structure by more than twenty-five (25) feet. If the antenna is to be mounted on a pre-existing structure, a full site plan shall not be required.

Well, it is pretty obvious that the township is steering telecommunication companies to existing structures. The special engineering involved precludes co-location and endangers the routine maintenance of antennas and cables not on safer towers. In addition, the locals have their hands out for outrageous rents and land for shelters might also be difficult to negotiate. This seemingly innocent policy gets the job done but hardly will provide for future site development. It is a short path to solving a single problem. I wonder how the smoke stack ever got approved? However, once your tower is up it looks like you will never have to come back to the zoning board.

A telecommunications site with antenna that is either not mounted on an existing structure or is more than twenty-five (25) feet (*I've often wondered how they come up with 25 feet)* higher than the structure on which it is mounted is permitted in all zoning districts by special exception.

All other uses ancillary to the antenna and associated equipment (including a business office, maintenance depot, vehicle storage, etc.) are prohibited from the telecommunications site unless otherwise permitted in the zoning district in which the telecommunications site is located.

Standards of Approval or Special Exceptions:

The telecommunication applicant is required to demonstrate, using technological evidence, *which will not be understood at all by most board members,* that the antenna must go where it is proposed in order to satisfy it's function in the company's grid system.

What if YOU are the applicant and not the PCS? Why can't you put up a tower with no promise that you will ever *get a client? Maybe you just like towers!*

If the telecommunications applicant proposes to build a tower (as opposed to mounting the antenna on an existing structure), *there is that existing structure bias again* it is required to demonstrate that it contacted the owners of tall structures within a one-fourth (1/4) mile *(1320 feet or 440 yards)* radius of the site proposed, asked for permission to install the antenna on those structures and was denied for reasons other than economic ones. *Can you believe that?* This would include smoke stacks, water towers, tall buildings, antenna support structures of other telecommunications companies, other communications towers (fire, police, etc.), and other tall structures. The zoning hearing board may deny any application to construct a new tower if the applicant has not made a good faith effort to mount the antenna on an existing structure.

There are some serious restraint of trade issues presented here especially for a tower operator who is in fact not a transmitter of rf energy.

Standards of Approval of all Telecommunications Antennas:

Antenna height. The applicant shall demonstrate that the antenna is the minimum height required to function satisfactorily. *This statement makes no sense as an antenna height will always be at the height for it to function satisfactorily.* No antenna that is taller than this minimum shall be approved. *This standard also contradicts a previous regulation for using existing structures. Antenna height is a function of many factors which might change or need to be changed at a later date. A tower provider selling "space" needs to provide various levels of access for clients "as inventory".*

Setbacks from base of antenna support structure. If a new antenna support structure is constructed (as opposed to mounting the antenna on an exiting structure), *Come on guys cut it out!* the minimum distances between the base of the support structure **or** any guy-wire anchors and any property line right-of-way line shall be the <u>largest</u> of the following:

(1) Thirty percent (30%) of the antenna height.

(2) The minimum front yard setback in underling zoning district.

(3) Forty (40) feet.

Antenna support structure safety. The applicant shall demonstrate that the proposed antenna and support structure are safe and the surrounding areas will not be negatively affected by support structure failure, falling ice, or other debris. The applicant shall also demonstrate that electromagnetic fields and radio frequency interferences comply with Federal Communications Regulations concerning such emissions.

We dodged a bullet here, what they really mean is "non-ionizing radiation" not electromagnetic fields. All support structures shall be fitted with anti-climbing devices, *razor wire works pretty well* as approved by the manufacturers.

Fencing. A fence shall be required around the antenna support structure, accessory buildings(s), and other equipment <u>unless the antenna is mounted on an exiting structure</u>. *Gees!* The fence shall be a minimum of six (6) feet in height and a maximum of eight (8) feet in height, shall completely enclose the antenna, *I wonder how that works since the antenna is at the top of the tower* support structure, and related facilities, shall not contain openings greater than nine (9) square inches and shall contain at all entrances, gates which shall be kept locked except <u>during such times as the site is manned or by authorized operations or maintenance personnel.</u> *Sure glad they thought of this one.*

Landscaping. The following landscaping shall be required to screen as much of the support structure as possible, the fence surrounding the support structure, and any other ground-level features (such as a building) *and the ugly fence they made us put up in the paragraph above* and in general <u>soften</u> *please define soften* the appearance of the telecommunication site. The township may permit any combination of exiting vegetation, topography, walls, decorative fence *flat surfaces with local artists using spray cans?* or other structures instead of landscaping if they achieve the same degree of screening as the required landscaping *which was never defined in the first place.* If the antenna is mounted on an existing structure and other equipment is housed inside an existing structure, landscaping shall not be required. *Bet you predicated that one huh?*

(1) An evergreen screen shall be required to surround the site. The screen can be either a hedge [planted three (3) feet on center maximum] or a row of evergreen trees [planted ten (10) feet on center maximum]. The evergreen screen shall be a minimum height of six (6) feet at planting and shall grow to a minimum of fifteen feet at maturity. *I sure have seen a lot of dead bushes around tower sites because with the big slab of cement forming the tower base, all the crushed rock and back fill, and the chemical ground material, as well as the rain water runoff from the hill the tower is mounted on, a bush hasn't got much of a chance.*

(2) In addition, existing vegetation on and around the site shall be preserved to the greatest extent possible. *Poison ivy, scrub and chic weed.*

In order to reduce the number of antenna support structures in the community in the future, the proposed support structure shall be required to accommodate other users *if paragraph "a" above is waived and not for free,* including other telecommunications companies, and local police, fire, and ambulances companies.

The telecommunications company must demonstrate that it is licensed by the Federal Communications Commission. *The only way to deal with this regulation is for the tower developer to go get a business band license for the site or have another licensee designate the site as a "dispatch point" naming the tower developer. This is regulation is way out of line!*

Required parking. If the telecommunication site is fully

automated, *perhaps what is meant is unattended?* adequate parking shall be required for maintenance workers. If the site is not automated, the number of required parking spaces shall equal the number of people on the largest shift.

Antenna support structures under two hundred (200) feet shall be painted silver or have a galvanized finish retained, in order to reduce visual impact. Support structures may be painted green up to the height of nearby trees. *Oh that's so beautiful!* Support structures two hundred feet (200) in height or taller shall meet all Federal Aviation Administration (FAA) regulations.

They forgot about the FCC Part 17 Regulations, proximity to airports, glide paths and that support structures under 200 feet could still be over 200 feet when the antenna is mounted on top. It would be better not to regulate this area at all in the ordinance as it is clearly not under the township's jurisdiction including the painting of the tower.

A full site plan shall be required for all <u>cell</u> sites *Good, we are putting up a PCS site so clearly we won't have to comply with this requirement* showing the antenna, antenna support structure, building, fencing, buffering, *buffering?* access, and <u>all other</u> requirements of the township subdivision and land development ordinances. **The site plan shall not be required if the antenna is to be mounted on an exiting structure!!!!**

<u>General Repeal:</u>

All ordinances or parts of ordinances, together with any amendments and supplements thereto, are repealed to the extent necessary to give this article full force and effect.

The provisions of this ordinance are severable. If any sentence, clause, or section of this ordinance is for any reason found to be unconstitutional, illegal or invalid *how about just illogical* such decision *who makes the decision?* shall not effect the validity of any of the remaining provisions of this ordinance. It is hereby declared as a legislative intent that this ordinance would have been adopted had such unconstitutional, illegal, or invalid provision had not been included herein.

The above example ordinance is in place. Many of the top telecommunication firms in the country today have grappled with it. Some complying others moving their proposals to another municipality. There is no question that the cynical comments rendered during its presentation here were meant to urge the reader to analyze every sentence and intent of the ordinance. Often the simplest statement can hang the whole process up causing months of delay.

Paragraph (l) requiring proof that you are an FCC licensee for example, could keep a tower owner/developer from building any tower in this township. Why then do they not require the owner of a smoke stack, water tower or other tall structure to likewise be licensed? Was it an oversight or intentional? The nauseating overstatement bias towards the use of existing structures may be an attempt to intimidate the communication company into signing with the first reasonable landlord (voter), not to mention his *out stretched hands.* On the other hand, what if you had the only tower in the township with excess capacity. Or perhaps looking at it another way, what if the township had the only tower in the township with excess capacity?

Hearing on phone tower flap rescheduled

FIG. 8.6 ONE TECHNIQUE IS TO DELAY ACTION INDEFINITELY

How typical is the above ordinance? As they go, this one is not that restrictive even though there are more than sufficient arbitrary and capricious points upon which a refusal to grant your petition for authority to construct a tower may be based. The purpose of this exercise is to stress the importance of your obtaining a copy of the ordinance with which you will have to contend, read it carefully and spin each paragraph to its logical (or illogical) conclusion. Be ready in you submissions to deal with ordinance issues not necessarily tower issues.

Finally, it is important to obtain a conclusion to your request for an approved zoning and building permit. Hopefully it will be for a similar project in your municipality. Typically final actions can vary from a one sentence letter to a complex *report and order* citing all the facts, testimony, objections and court cases going back to the Inquisition. The problem with this sort of order is that it takes months to write, rewrite and usually won't fly on the first release requiring several appeals before construction will be authorized. Let me give you an example:

> "Since the classification of a particular use as a special exception use evidences a legislative finding that the use is consistent with the zoning plan, a properly supported special exception request should only be denied where the adverse impact upon the public interest exceeds that which might be expected in normal circumstances. See Brentwood Borough v. Cooper, 60 Pa. Commonwealth Ct. 462, 431 A.2d 1177 (1981); Pennypack Manor Nursing Home, Inc. v Petrella, 35 Pa. Commonwealth Ct. 367, 371, 387 A.2d 139 141 (1978) (holding that to be the rule in cases involving special exceptions). Mere possibility of adverse impact is not enough. See Brentwood Borough v. Cooper, 60 Pa. Commonwealth Ct. 462,431 A.2d 1177 (1981); Warren County Probation Association v. Warren County Zoning Hearing Board, 50 Pa. Commonwealth Ct. 486, 489, 414 A.2d 398, 399 (1980). Nor are generalized condemnations fears sufficient to prevent issuance of a special exception. Loch v. Zoning Hearing Bd. of Borough of Emmaus, 131 Pa. Commonwealth Ct. 211, 569 A2.d 1035 (1990)"

Site Savvy

> Often you will need to deal with *ordinance* issues rather than *tower* issues

The above paragraph taken from an actual appeal of a communication company's initial rejection for the placement of a 120 foot monopole addresses only the board's own decision of allowing an exception to its *conditional use* list for the erection a tower within an a L1-light industrial zoning district. That's good! The fact that the tower was to be located next to a pretty dirty cement plant was of no concern. Hopefully your application for a building permit will not be met with the kind of opposition suggested in this section. But on the other hand, if you are able to perceiver, you might wind up being the only ball game in town and have a real competitive edge over newer tower entrants.

FIG. 8.7 POLE TOWER PROVIDING CO-LOCATION OPPORTUNITIES

8.7 Other Siting Strategies

Conversion of exiting structures to antenna support systems can be an interesting alternative to the head banging approach of trying to get approval for a tower. "When zoning people reject the steel, you bring out the wood" says Don Simon of Mountain States

Marketing in Loveland, Colorado. Laminated wooden poles in ski areas look a lot like power poles and are not easily identified as even being PCS antennas. They can be placed directly in the ground in a drilled hole and can usually survive ice and a wind velocity up to 130 mph. The cost is a lot less also. Laminated towers can look like an observation deck or carry some rustic charm and may represent an interesting alternative to steel. The company can design artistic shapes up to about 120 feet in height.

Billboards also represent an interesting location for antennas often with no zoning problems at all. This area has been pretty much developed throughout the nation by billboard companies, and antenna management firms representing them. More information can be obtain from the Outdoor Advertising Association. So if you own land along the highway and are being solicited for a billboard, cut a deal to share in the siting revenues that may result, or build your own for a double dip.

Stealth sites cost a lot of money. But from time to time you might be in a position to capitalize on a location where a stealth situation will make sense. A monopole that is supposed to look like a pine tree can actually look pretty stupid. But flag poles, light standards, mock silos, phony water towers, a bell tower, a church steeple, a faux chimney (a non-smoking stack) and fake wind mills can be pretty nifty alternatives to a hard nosed zoning board. Check with *Stealth Network Technologies Inc.*

FIG. 8.71 A TOWER MADE TO LOOK LIKE A TREE MAY DRAW MORE ATTENTION THAN JUST A TOWER

8.8 FCC Preemption of Zoning Board Regulation

The headlines read **"Industry Lambastes FCC For Not Dealing With Siting Moratoria"** by zoning boards. A moratoria is another way to halt the approval of towers in municipalities until zoning ordinances can be put in place or until a "deal" can be struck or until a problem goes away. It has a chilling effect of stopping dead in its tracks the billions of dollars invested in the *free entry* of new and competitive telecommunication services. A lot of communication companies, tower investors and operators are tired of this process and have placed their appeals and hopes with the FCC to do what they said they were going to do. That is to preempt local regulations.

But the matter is not that simple at all. The issues of harmful non-ionizing radiation can't seem to be resolved to anyone's satisfaction. Issues of aesthetics are totally arbitrary. The whole process might even be driven by factors which are not evident in the ordinances themselves. Has it all happened before? Cable companies had to deal with *franchises* before they could wire up local municipalities some of whom get a piece of the action. Then in another breath the same municipality restricts the *unsightly* installation of satellite dishes negating the free and alternative choice by the public for a cable alternative. A new service is created and DSS dishes are permitted, but attempts to have their *installation* subjugated to local control by some municipalities was squashed by the industry quickly. Here the FCC came to the rescue with a preemption of local zoning for new DSS dishes under a certain size.

Could a revenue stream to local government be one reason for controlling towers owned by anybody other than the municipality, all in the name of looking out for the public good? One thing for sure, it will take many more years before the real motives are clear. The communication industry trade journals are an outstanding source of current information for what is going on in all these issues. Be sure to keep up to date with each story which is reported. Your mission however is to discover what has to be done to get **your** tower approved so that you can build it as quickly as possible before the opportunity window closes forever.

An extreme ongoing case involves the city planners of McLean, Va. vs. AT&T Wireless Services. This case began with objections levied by the wives of Zbigniew Brezinski former National Security Advisor and Ed Meese, former Attorney General under President Reagan. AT&T wanted to build a crucial antenna on a 3/4acre plot *zoned commercial* across the street from a gas station. Dubbed the *Eaton* site, it was one of seventy-five sites being constructed in the company's new PCS system serving the DC and Northern Virginia area.

Nearby, a concerned neighborhood for "affluent professionals" had homes ranging in value in the six to seven figure range. This case was similar to the one in Medina, WA, also near some *high end* residential property of folks like Bill Gates, whose community successfully frustrated attempts by Sprint PCS to build a tower there also.

It's not like AT&T didn't do everything it could possibly do to obtain approval. It conducted three-hour seminars *"Reach Out and Touch the Future";* where citizens and government officials had an opportunity to discuss the issues with company officials and engineers.

AT&T even designed its tower to look like a *pine tree* which they said would cost four times what an uncamouflaged monopole would run. They also provided for *co-location* in the event another carrier could use the site.

The issue raged for over a year. In a nearby county, AT&T obtained immediate approval for a similar 180' monopole with no problems. The Fairfax County appeal (where McLean is located) centered upon alleged *failure* by the county *to document its rejection of the site*.

This case may not be over. But it points out the necessity to draw a compromise (wealthy or influential personalities notwithstanding) between the population interests and the carrier's interests as well as county responsibility, in order to "balance private needs with community objectives". But certain facts remain. The tower could not be located on county owned land (steering ?) and was needed to effectively operate a duly FCC licensed system.

Perhaps the solution to all these problems is a new technology in antenna design. If you ever have seen a monopole that looks like a tree its pretty hard to keep from laughing, especially in the winter. It also seems to me that it memorializes the ridiculousness of a "process" that puts it there in the first place.

If you discover any future development for *high end* residential homes currently being planned for the future in your area, build a tower nearby immediately! Then sit back and smile.

CHAPTER NINE

9.1 Government Site Rules and Regulations

There are many government rules and regulations required in the operation of an antenna site. These are not the rules that you presented to your tenants in the previous chapter even though some Federal Regulations might be reflected in your "house" rules in order to shield yourself from liability. Federal Regulations covering towers and roof-tops are essentially promulgated by the Federal Communications Commission even though you might find other Federal agencies involved such as the Federal Aviation Agency and OSHA. In addition, there are state and local regulations which need to be considered from time to time depending upon your area of operation.

We will be discussing three (3) basic areas of Federal Regulation for site owners:

1. RF Radiation Safety Regulations.
2. General Tower Requirements and Registration.
3. Aviation Flight Hazard Regulations.

9.2 National Organizations Can Assist

There are a few national trade organizations dedicated to antenna site owners, managers and operators who can assist in the compliance and lobbying efforts related to rules:

Personal Communications Industry Association's
Site Owners and Managers Association SOMA
500 Montgomery Street
Suite 700
Alexandria, Virginia 22314
800-326-8638 Voice
703-836-1608 FAX

This organization is part of PCIA consisting of a membership *who manage, own or operate transmission facilities and sites for wireless telecommunications or who are engaged in site acquisition, construction, or servicing of sites and facilities for wireless telecommunications.* SOMA provides a common voice for national site issues. In addition, SOMA has created several publications of interest such as <u>The PCIA Co-Location</u>

Guide for identifying companies and contacts in the antenna site industry. The publication is listed by state and is intended to be a resource for "helping wireless communications service providers implement a co-location strategy by identifying primary professional site management companies offering properties, sites, towers or structures suitable for mounting antennas, transmitters, and base stations."

Another publication of interest is Communications Towers Connecting Communities-Connecting People. This handout style brochure is sold in lots of 100 for about .75 cents each. It is an attempt to *cut through some of the misinformation and incendiary rhetoric sometimes spread by groups opposed to new structures*. A bit busy and hard to follow, the brochure is a good start in the *educational process* which the organized siting industry believes it is under some mandate to propagate. There is also a thirteen 13-page booklet which *provides an overview of wireless communications and its infrastructures*. The booklet runs about $4 for non-members. Finally, the SOMA Professional Standards Committee has completed a SOMA Tower Manual. This publication consists of a series of articles (not intended to be industry standards) and general information relating to technical and operational matters along with a compendium of recommended practices which could be useful to site managers. It was also designed for site customers to evaluate the relative quality of service offered by different site managers. Check the PCIA Web page at **http://www.pcia.com** for a listing and prices for all the current publications offered by this organization. As always an organization of this type typically has two rates, one for members and the other for non-members. Some of the publications are rather expensive and their value may be questionable to smaller operators. Fortunately there are some rather good summaries written about some of the publications which can be found on the PCIA Web page. Contacting PCIA directly will also allow you to obtain information on membership and other services offered.

FIG 9.2 PCIA BROCHURE

Another organization is the National Association of Tower Erectors, NATE. This organization, although primarily directed toward contractors and tower manufacturers offers an excellent monthly publication called *Tower Times*.

National Association of Tower Erectors *(NATE)*
420 4th. Street NE
Suite 109
Watertown, South Dakota 57201
605-882-5865 Voice
605-886-5184 FAX
http://www.daknet.com/nate

The organization has several committees working in areas of training, trade shows, OSHA, construction and safety. Articles in the *Tower Times* deal with many issues of concern to the siting industry, are well written, and are usually comprehensive how-to in nature. In addition, advertisers have an opportunity to present technically interesting advertisements. Another advantage of this organization is that the NATE Insurance Programs could provide some insurance alternatives for members at reduced cost. As trade associations go, this one is at the top of my list and is worthy of your consideration.

9.3 RF Radiation Issues

In August of 1996, the Federal Communications Commission issued a Report & Order on *human exposure to radio frequency radiation*. The release suggests that over 100,000 antenna locations (per year) will need to be evaluated as to whether or not these antenna locations comply with the new regulations for *radiation* exposure. The *effect* of this regulation release may have triggered fears in advertisements and other claims such as mental illness, impotence, cancer and a host of medical problems associated with the non-specific generation of RF energy. The interpretation of these engineering guidelines were met with additional charges of *what really are the hidden effects* of RF and MW energy? It is what the Report & Order didn't say that seems to have captured the minds of environmental activists almost immediately resulting in efforts to oppose the installation of telecommunication towers. The issues have been around for decades. Perhaps it was an inopportune time to release such an Order at the same time that so many towers and antenna sites were needed in the emerging wireless infrastructure.

The actual release however is pretty technical and a match for even the best high school math student. The industry even had a tough time coming up with a common name for the radiation involved. It has been called Radio Frequency Radiation (RFR); Electromagnetic Radiation (EMR); Electromagnetic Energy (EME); Electromagnetic Fields (EMF);

Microwave Energy (MWE); Root Mean Squared alternating current (RMS); and perhaps the most accurate description, Non-Ionizing Radiation (NIR). As of October 15th. 1997 the FCC is in charge of the new law.

At one time the FCC wasn't too concerned with the low power typical for most cellular and PCS operations and did not require the issues of potential human exposure to radio frequency emissions at these antenna sites to be addressed. But based upon new data and the fact that antenna sites, particularly roof-tops, were loading up with transmitters and were nearer to humans (in the top floors), most wireless operators are no longer excluded from analyzing their sites especially where the *cumulative effects of many transmitters* might be involved. The new Rules (based upon OST Bulletin Number 65) indicated that licensees operating transmitters that contributed more than 1% to the total power density (more practically 5% of an exposure standard) *are responsible* to insure *compliance* in areas that exceed the exposure limits. Licensees now need to comply with the Maximum Permissible Exposure (MPE) for RF radiation.

These exposure limits are measurable both by theoretical methods and by instrumentation.

A. The theoretical method is complex requiring knowledge of the actual antenna patterns, radiation lobes, power levels and frequencies. However the problem is being addressed effectively with the introduction of specialized software such as *RoofView*™ or *TowerView*™ created by Richard Tell Associates, Inc. **rictell@accessnv.com** . These programs simplify the process of both calculating and presenting the systems data in a multiple or complex RF site environment. However the program is only as good as the data which is input and the process takes time and is subject to many assumptions.

B. The instrumentation method, fortunately takes all of this into account and presents usable data both as a printout in addition to presenting a meter reading which instantly converts all the radio frequency (RF) energy being radiated (at the time) into a compliant determining percentage exposure definition.

Simply stated, **Non-Ionizing Radiation** (NIR) heats the skin. Ionizing radiation, like X-rays and Radon, can actually destroy living tissue because their effects are cumulative over much longer periods of time. RF produces Non-Ionizing Radiation which can be predicted and directly converted to the heating caused to a square centimeter of exposed human skin. In addition to exposure intensity, *time* of exposure standards have been set at six (6) minutes for a *controlled environment* and thirty (30) minutes for an *uncontrolled area*. The frequency of the radiation is also a factor altering the parameters somewhat over several frequency bands. For example, from .3 to 3 MHz., the frequency range in which the AM Broadcast Band is located, an electric field strength of 614 V/m has been set along with the corresponding magnetic field strength of 1.63 A/m. The electric and magnetic fields are but two (2) of the components of a typical RF signal. In addition, there is a component called the Power Density or *plane wave equivalent power density* set at a limit of 100 mW/cm2.

What all this means is that measurements or calculations must address both the electric and magnetic field planes at lower frequencies but become less important as the frequency rises, say up to the 300-1500 MHz. band where ESMR, cellular and some paging services are located. The FCC report contains complete details of these parameters and is available for downloading from the FCC Web site. **http://www.fcc.gov/**

Interpretation of the actual operating data is our objective. Also we must have an understanding of how the information is used. Co-location and antenna positioning, now in populated areas as opposed to remote mountain top locations, is the reason for the emphasis upon these Rules. The Rules were developed from hundreds of studies, reports and papers over the decades, a majority of which have been listed in the FCC bibliographies. Many theories have been presented. Accordingly it is not surprising that various standards have been developed and presented in a number of differing ways. Very much like tower loading parameters, there can be many alternative definitions essentially explaining the same phenomena adding to the confusion of just what terms to use, what formulas to use and what the results actually say.

For instance, in January 1996, Clark W. Heath Jr., MD, wrote an article for the <u>CA Cancer Journal for Clinicians (Vol 46 No.1 pg. 29-44):</u>

The idea that exposure to power-frequency electric and magnetic fields (commonly termed "electromagnetic fields" or EMF) might contribute to human cancer causation has been under close study for nearly two decades. While various hypotheses have been advanced to suggest how such exposure might promote carcinogenesis, the evidence remains inconclusive. Data are of two kinds: (1) epidemiologic observations concerning the occurrence of childhood and adult cancers under different residential and occupational exposure conditions and (2) experimental studies seeking reproducible evidence of carcinogenic effects arising from EMF exposure in cells, tissues, and laboratory animals.

To date, no form of electromagnetic energy at frequency levels below those of Ionizing Radiation (x-rays) and ultraviolet radiation (our Sun) has been shown to cause cancer. Nonetheless, public interest has been vocal and continuous in zoning meetings because of the presence of electrical power and other highly complicated and seemingly mystifying technologies now in use by society.

The telephone has been around for over one hundred years. Its ear piece functions as a result of an electro magnet creating magnetic fields which control a diaphragm converting electrical pulses to mechanical energy and subsequently to sound energy. With the billions of hours this *electromagnetic generator* is compressed to the human skull (ear) is it any wonder that someone hasn't proposed that this energy is responsible for mental disorder, depression, anxiety or other maladies. Not considering that what might be discussed on the phone could be a greater causative agent for stress, are you willing to bet that this issue will not someday be brought to the point of being of concern to the health of the population?

Dr. Clark concludes:

While it may be impossible to prove either the presence or absence of risk, perhaps it can be feasible to assign likely ***risk boundaries*** *upon which practical guidance for community consensus can be reached.*

The FCC Report & Order clearly establishes these risk boundaries as guidelines for the present and offers the best policy until other more credible data can be presented and proved effective.

There is an organization called the EMR Alliance, an international alliance of grassroots groups whose primary concern is the adverse effects of EM, RF and MW radiation. They have a quarterly journal called *Network News*. **emrall@aol.com** The organization seeks to arouse public concern about health and environmental issues *not because they are anti-industry, wide-eyed activists or rabid environmentalists as the wireless industry is so fond of saying,* states Cathy Bergman-Venezia co-author of the *Network News*. Citing studies in the United Kingdom and Australia, any increase in leukemia incidence was associated solely to a close proximity to TV or FM broadcast towers. Apparently this is evidence *enough to justify tower construction moratoriums before zoning authorities.* Moratoriums were indeed imposed demonstrating that fears are sufficient to *stand up to a billion dollar industry*. The issue may never be resolved to everyone's satisfaction. In a recent German study the conclusion was drawn that an increase in blood pressure from 5 to 10 mm Hg. was "probably" due to the constriction if arteries caused by RF radiation given off from a cell phone. Probably? They are going to have to do better than that. But for the present, be prepared to defend your venture from this perspective. A good start is this concept of **risk boundaries** proposed by both *our* American medical studies and *our* own regulatory agencies. Remember the *inverse square law*. The further you get away from a radiating antenna, the faster the RF energy decreases. Try to use this principle for setting safety limits adopting the most accurate data available in the formulation of your policies if you are called upon to address these public safety issues. Now, what do the rules say?

9.4 RF Radiation Regulations

Strong *RF fields* have been around for about a century in the area of radio and fifty years with the advent of television but only recently in the industrial and production sectors. Here for example, industrial, scientific and medical frequencies use RF energy for the production of semiconductors (13.56 MHz.); RF heat sealers (27.12 MHz.); Industrial microwave drying equipment and consumer microwave ovens (2,450 MHz.). At last count there were over 70 patents on metal processing and vinyl molding and pressure injection systems using very high outputs of RF energy to create precisely controlled heating conditions. Many of these systems are in the 10 Kilowatt range with shielding that may be less than any standard providing for the protection of workers.

The human body has always been dependant upon narrow heat parameters. The body temperature, a nominal 98.6° Fahrenheit (F) can have variations as little as .2° F. Such changes can indicate infection or just being out in the Sun too long. Human exposure to RF energy has been subject to standards for years with the IEEE C95.1-1992 and National Council on Radiation Protection, NCRP Report 86, suggesting controls for exposure levels to which an individual human might be subject.

The key to understanding the Rules however is the difference between *controlled* and *uncontrolled* exposure. Supposing that a large co-located site survey has revealed a specifically high RF level reading, perhaps at a position half way up the tower. A typical high level reading could be where the *Power Density* reads greater than f/300 mW/cm2. This would indicate that climbing into or near this area would result in predicted <u>skin heating</u> which could be dangerous if the period of the exposure were to exceed a defined six (6) minute duration. The problem is obvious, as you get closer and closer to the source of this radiation its levels are going to rise beyond a *safe* limit. Thus your body temperature will begin to rise.

The situation is made practical by a service technician who must change a beacon lightbulb at the top of a television transmission tower. There could be a point where the <u>lower lights</u> could be changed within the required safety limit, but as the climber gets closer and closer to the radiating elements of the antenna at the top of the tower the limits are exceeded. The technician's body temperature is going to elevate and at some level there may even be a point where dizziness or unconsciousness could result. What to do in this case? Turn the TV transmitter off until the required maintenance (or more realistically transmit from a back up site) can be completed. These procedures have been in place for years and there is nothing new here. What is new, is that the site must be marked as a *controlled* area with specific signs and fall under a set of predefined procedures which regulate the operation of access for **all personnel** at the site not just technicians. Supposing unsuspecting grounds people, just mowing the grass are subject to increased body temperature? Once again the fields are measured, fences (access) are then moved *out to the threshold of the defined radiation*, signs are posted and rules of access are imposed.

Fig. 9.4 A sign marking a *controlled* area

Fortunately, in addition to instrumentation to measure these fields, **body monitors** are also available. These little pager-like boxes are worn on the belt affording an immediate alert to the worker should RF fields come close, equal or exceed the defined limits. The key to proper implementation of an RF Radiation (RFR) standard is a written policy and an in-place procedure which has been formally accepted by your clients.

There are reasons why this policy is advisable. It not only insures compliance by and for your tenants, it provides the necessary justification to local zoning boards when they address your intent to add additional antennas. It also establishes an operating history of what is happening at your site. In the long run it will be a requirement to be routinely reported in subsequent license applications before the FCC.

If technicians working at a *controlled* site do not know what it means, then the whole process falls apart. So if you control the site, it falls upon you to see to it that workers, including visiting or temporary ones know what it means when they see the *sign*. For instance, you manage a roof-top which is a *controlled* area. The *signs* are posted. An air-conditioning crew who know absolutely nothing about RFR is called to repair a unit near active antennas. Contemplate for a moment how these service people are going to know what to do or even if they are in a *hazardous* area. That is your problem. Most communication companies now have corporate policies implemented at the highest level with training programs for workers in the field. The training consists of a short video tape explaining the potential hazzard and the remedial and monitoring methods which must be followed. Much of this policy can be posted at the site with maps showing the various average levels and explaining the potential for hazzard in addition to telephone numbers for various engineering personnel should facilities need to be deactivated for maintenance or other work in areas where over exposure is likely. Without an organized approach, these matters can elevate to extremes and can cause legal and public relations problems.

FIG. 9.4.1 NON TECHNICAL PERSONNEL HAVE NO KNOWLEDGE OF RFR HAZARDS AND MUST BE INFORMED

From a technical point of view, both electrical and magnetic fields are more complex than just what is present around the base or *near* antennas. Referred to as *near field* and *far field* components, various sets of data must be explored and analyzed in terms of their potential for human exposure. FCC guidelines provide approximate models that may be used to calculate power density in various frequency bands as well as the patterns of radiation developed from the specific design for the actual antenna used. PCS and Cellular antennas can be both *omnidirectional* and highly *directional* and *push* RF energy out in narrow beams perpendicular to the side of the antenna. A pattern can have both horizontal and vertical components. If one moves out about 150 feet from a typical cellular antenna, its beam or *far field* radiation is highest for the pattern developed by its physical design. Often this radiation is actually directed away from the mounting structure and that is good. But what about just under the antenna. What kind of energy levels reside there? Ideally there should be deep nulls in the directions of undesired energy, after all why beam energy straight up or down when the energy is not needed there? As ideal as these radiators have become and with ever improving designs, less and less *null fill* is evident in the real world with the advantage that *near field* radiation (typically around the base of the antenna, tower or roof-top) is minimized. Additional shielding can also help in reducing this component of measurable RF intensity. Nevertheless the majority of potential for RF hazard zones still exists near the antenna in the *near filed* zone especially if the antenna height is equal to where a human could stand and look directly into the vertical plane of the antenna. But both components must be observed, recorded and compared over time to insure complete understanding of what is taking place at your site. The two methods of measurement mentioned in the first part of this chapter should be understood fairly well in order to minimize inaccuracies in the collection of data. Even though software has been developed predicting fields surrounding antennas, the situation becomes much more complex when you attempt to combine all the effects of all the antennas operating simultaneously. Determining reliable data should be a function of your confidence in the consulting engineer who performs the analysis. But more often than not it will be left to you to assemble the raw data which will be used. The frequency, power levels, type of antenna, mounting location etc. are all practical data elements required in the final calculations.

Formulations for *far filed* and *near field* models are contained in the various releases by the FCC such as Evaluating Compliance with FCC-Specified Guidelines for Human Exposure to Radio Frequency Radiation, OST/OET Bulletin Number 65; and, from other publications such as Richard Tell's, Engineering Services for Measurement and Analysis of Radio-frequency (RF) Fields, Office of Engineering and Technology, FCC.

Another publication, Guidelines for Evaluating the effects of Radio-frequency Radiation FCC Report & Order, ET Docket No. 93-62, August 1, 1996 should also be obtained. All of these guidelines and additional material are available from the various Web pages presented in this section.

Methods for estimating power density in the *near field* for example can use the *cylindrical* method which involves the average power density on the surface of a cylinder, with a height equal to the antenna's aperture and a radius equal to the distance of interest. The model is useful in the *near field* within the aperture of the antenna. Another method called the *collinear* method can be used for an array of elements with a length of one half wavelength. It is possible to estimate the power density of this array in both the *near* and the *far field to* within one wavelength. The accuracy of this technique is dependant upon how well the array of linear elements, fed in phase, represents the real antenna. The *collinear* method estimates the number of elements in the array and then the gain and pattern for each element. The power density near the antenna is calculated by combining the contributions from each element in the array. Even though much has been written on these methods, when used with typical or similar antenna types, there are bound to be major differences. But it probably will not be long before antenna manufacturers will be providing a nomograph for each of their antennas just as they now do with published pattern and power gains. This will no doubt simplify the theoretical approach to calculating power density at various selected points around a radiating element. In order to have confidence in these models however, actual power density measurements are made and the predicted and actual calculations are compared. This suggests that perhaps the direct reading approach makes more sense and should be used instead of trying to predict on paper what the results will be. However it still may be necessary to use pro forma predictions if zoning boards want to know the "effects" *before* an installation. This would suggest that a theoretical determination will still have a place in the calculations for a time.

My experience has been with the second method, i.e. actually measuring the RF and magnetic fields with a handheld instrument. Relying on digital circuitry and an uniquely designed RF probe, actual readings may be obtained over defined frequency ranges on a bar graph display with instant averaging of various bands. The cumulative effects of **all** operating facilities are computed and presented both on an analog meter as well as detailed printouts (when the instrument is plugged into a computer and interrogated). With the meter, you are instantly presented with a compliance result. But more importantly the instrument can locate all sorts of problems that the theoretical method will never find.

For example. In doing a survey using the NARDA Model #8718 at a very congested tower site, a quick scan of the transmitters in the building, not usually a part of the RF compliance procedure, produced three interesting observations. The first one involved a coax connector that was loose connecting the antenna transmission lead to a 500 watt transmitter. The loose connection was literally *spewing* RF all over the equipment room. The last technician working on his system apparently did not tighten the connector after servicing. The performance of the transmitter was probably degraded as well. Merely tightening the connector eliminated the RF level in the equipment room as evidenced by an immediate drop in the meter's output reading. Finding the connector was easy. A frequency readout indicated the offending energy was near 900 MHz. The signal strength indicator was followed as it increased until the actual connector was pin pointed.

In another instance the ground tie bar going into the building appeared to have an abnormally high RF leakage at a broad range of frequencies. A check of the ground wires indicated that one of them had been sheared where it connected the buss to the ring ground (probably by a power lawn mower). A repair was made and the ground buss returned to it's normally low RF output level.

Finally, a coaxial junction box in a TV transmitter room was very "hot" and found not to be in compliance when the probe was placed within a few inches of the device. The system was the point where dry nitrogen gas was inserted into the transmission line. This is not an area that would receive much attention for RF contamination problems. Some cleaning and reseating of the brackets brought the RF levels down to below detectability and the system was returned to normal.

Fig. 9.42 NARDA Model 8718 monitor for RFR energy

These three problems were solved inside of an hour and weren't even suspected nor could have been discovered had not an instrumentation survey been done at the site in addition to the RF levels measured for the purpose of controlling and defining the RF at the site. A few hours with the "8718" solved more problems than were even envisioned when the survey was begun.

In practice the digital display portion of this instrument and others like it is the smallest contributor to measurement uncertainty, stated Richard Strickland, Director of Business Development for NARDA

Microwave, Hauppauge, NY. The instrument (similar to the Model 8718) is essentially a hand held meter to which is connected several very specially designed probes. One probe measures H Field (EMF energies from power lines for example) and another measures the E Field (RF measuring probe). These probes are in the $2500 price range. The basic meter runs about $3500. Other less complicated models are available. Because the probes have a *shaped* frequency response designed to correspond with the standards as defined, they must be treated with extreme physical care and calibrated frequently (at the factory) to insure accuracy. The real advantage of direct instrumentation however is that an actual reading at every point of interest is immediately available showing not only compliance (or non compliance) but when corrections have been made the results of those changes are immediately available proving the validity of the remedial action. Together with a formal RF Radiation Safety Program, regulatory compliance can be confidently and conclusively demonstrated to not only tenants, who may not suspect their contribution to the problem until you show them, but to local government officials and the public in terms that are easy to comprehend.

Site Savvy

> The FCC has agreed that if appropriate and effective action is taken to ensure workers or members of the public are not exposed to RF fields in excess of the guidelines then the facility would be considered to be in compliance.

9.5 Proposed Policy Statement

The following draft policy statement creates a Radio Frequency Radiation Safety Program ("RFR" Safety) and should be used as the basis for establishing a company *program of record* showing the intent and a procedure for the implementation of a safe environment for workers.

To: All personnel, Antenna Site Tenants and Subcontractors.

The Federal Communications Commission and the Telecommunications Act of 1996 now specifies that all antenna site owners and operators provide a safe and hazard free environment with respect to Radio Frequency Radiation (RFR), also known as non-ionizing radiation. Antenna sites such as roof tops and common use towers now require both administrative and engineering RFR hazard controls. Equally important to our efforts is an awareness training program for our employees, our telecommunication clients and in some cases their subcontractors who work in areas we control where sources of RFR might be encountered.

Our company is likewise required by the FCC to include similar certifications of compliance with federal guidelines concerning human exposure to RFR in license renewal applications.

Accordingly, I am establishing a formal RFR Safety Program under the direction of____, who will establish specific policies, procedures and a training program for our employees. In addition he (or she) is authorized to establish a Safety Committee to assist in this important activity. The Committee will make recommendations concerning implementation practices; required acquisition, distribution, operation and calibration of approved monitoring equipment; will schedule site studies, coordinate the posting of signs and provide for the proper notification of findings for all our installations.

Our objective in establishing this policy is to meet federal, state, local and zoning board requirements for a controlled RF environment so that we may continue to operate and safely expand our maximum permissible exposure (MPE) limitation as specified in the FCC Rules. Questions concerning this policy may be directed to___.

After the policy is issued you will need to prepare a list of procedures and practices. Since this area depends pretty much on the type of site you will be operating, it is suggested that you contact similar tower companies requesting copies of their procedure. There are also excellent published articles covering this subject as well as specific consultants who can assist in the development of a program.

9.6 RFR Training Programs

At the time of this publication there was an excellent one day Public Seminar circulating the country as well as a detailed three day Course in RFR Safety Training. Call the NARDA Seminar Center at (516) 243-0838 for your scheduled public seminar brochure. The program is well worth the effort. In addition, the NARDA catalog is a must have product guide explaining many of the complicated issues and product offerings of the company. Write to Loral Microwave-NARDA, 435 Moreland Road, Hauppauge, NY 11788.

There will be many adjustments to the RFR standards and policies but eventually rational information and the body of knowledge will prevail. The subject is worthy of your continued attention and consideration for there could be a potential for income in your understanding of field survey work in this area as a continuing service to the telecommunication community.

Finally a word about an overstated RFR issue . . . pacemakers vs. cellular telephones. The image of a doctor making a cellular call in his office and a patient in the next room going into cardiac arrest just isn't happening. A Washington based consulting firm, *Wireless Technology Research,* surveyed over 1000 *pacemaker patients* placing operating wireless telephones right next to the chest. What was found was that while half of the heart-regulating devices experienced interference from cellular phone use, only 7 percent of the patients actually experienced any symptoms from the RF interference. Fewer than two people in 100 (or 1.7 percent of those pacemaker wearers in the study) actually experienced symptoms strong enough to cause concern about their health. Remember, the distance for this study was three 3" inches from the pacemaker a distance that would not permit practical use of the device by the patient.

The final recommendation? People wearing pacemakers should keep their cellular phones *off* when wearing them in a *shirt or jacket pocket* anywhere close to the chest. Companies testing pacemakers should test and include improved RF filters in their products. But here is some news. It seems that the technology used makes a difference. Analog cellular service caused less than 2.5% of the problem, while TDMA type phones caused most of the problems at 52.6%. The new PCS1900 service caused less than 1.9% of the interference problems.

9.7 FCC & FAA Antenna Site Considerations

Another area of site regulation involves a Report and Order in WT Docket 95-5 released November 30, 1995. The FCC adopted rules to streamline the Commission's antenna structure clearance process by requiring owners to register certain antenna structures with the FCC. The registration requirement generally applies to structures more than two-hundred (200') feet above ground or located near an airport. This Rule does not alter any of the procedures for proposed antenna structures or existing structures that do not require a new aeronautical study by the FAA. However proposed antenna structures must be registered prior to construction and existing antenna structures that do not require a new study must also be registered.

Since the program of tower registration before the FCC is relatively new, an initial filing schedule (by state and date) was established to record existing structures while distributing the work flow and application burden during the process. Now that the scheduled filing for all structures is over (June 1998) routine filings should be made for any new sites, modifications of existing sites or data corrections as they are needed.

As a bit of background, the Federal Aviation Administration (FAA) administered its tower and high building flight hazzard approvals through published regulations of its own. FAA Part 77, is concerned with tower height, lighting, proximity to airports and other general flight safety issues. Applications are made to the FAA for a *determination* as to whether or not the proposed antenna (structure) would be considered a hazzard to flight operations. The FCC on the other hand, in their applications for frequency use, required a general statement to the effect that the proposed antenna structure "had been analyzed" under the FAA rules. An FCC public service subdivision, the Antennas Survey Branch, then reviewed these submissions on a brief form which they had created. Needless to say the FCC seemed to be requiring that the *determination* be made prior to acceptance of a License Application. This meant that if several thousand FCC applications in a filing window were to be made, the FAA would be inundated with requests for "hazzard determinations" for sites that more than likely would never be constructed. Delays in processing thousands of applications by an FAA bureaucracy that was only able to review a few dozen applications per week ensued. Objections quickly followed by FCC filers. In many cases competitive FCC filings might be dismissed from consideration as technically defective because the *determinations* were not included in the original application as a result of not being received on a timely basis from the FAA. It was a typical governmental "grid-lock" and something had to be done.

There also emerged a few other problems of major proportions. For one thing many existing towers and the hundreds of FCC licensees "hanging" on them, were on site locations where the original Longitude and Latitude may have been based upon very old land survey techniques. Towers constructed in the 1940's and 1950's using these land based survey methods literally disappeared from the FCC's data base when the new satellite photos and NAD83 co-ordinate standards were applied to revised topographic maps and then used to impute co-ordinates for new licensees using the same facility. In earlier maps, tall *Radio Towers* were even located on the maps themselves. A licensee had two choices, use the coordinates published on everyone else's FCC License at the site along with any original FAA Flight Hazzard Study, or resubmit the paper work using the new maps. Both methods were used in some cases without the knowledge of the original tower owner. Neither party nor technique was wrong but the situation frequently created two (2) tower locations for the same facility. The problem really started to compound in the early 1980's when cellular service and paging facilities mushroomed creating thousands of new antenna sites and applications at existing locations. Licensees were no longer going to take the chance of improperly locating even an existing tower in spite of what was in the FAA data base.

The other apparent problem was of a procedural nature. Certifications on FCC forms asked if the antenna site had been analyzed under FAA Part 77 and if a study was required or not required. The assumption was that the **applicant was going to perform the analysis** based upon the Rules. With liability issues becoming more of a concern to the regulatory bodies, i.e. tower/plane crash incidents, any *informal* determination seemed to be foolhardy when a Government Agency (the FAA) was required to give you an *official determination* on an official government form as an alternative to just *your* opinion. After all, there were no fees for this request and having an official hazzard determination from the FAA just made good sense. When additional licensees showed up to use the tower they would merely copy the official determination and include the all important FAA Study Number on their FCC forms when the question was asked. Needless to say the FAA was not pleased with the extra paperwork especially for structures that clearly required no flight hazzard study by definition in the Rules. Yet if a formal request was made, they usually had to reply, if even in the negative, in accordance with their own procedures.

Alaska Tower/Plane Crash Claims Two Lives

by Lauren Rooney

NOME, Alaska The crash of a plane into a radio tower in Nome, Alaska took two lives and left KICY(AM) off the air.

The station was knocked off the air June 27, when a Cessna 207, belonging to charter service company Olson Air, hit its tower in dense fog. The crash occurred around 4:30 p.m. as the plane was heading for the Nome airport from Brevig Mission, Alaska.

"We don't know why the pilot was flying so low, other than to get below the fog as he approached the airport," said KICY General Manager John McBride. Witnesses said that as the plane approached the tower, the pilot gunned the engine. But the plane's right wing sliced into the tower about 35 - 40 feet from the top; both the plane and the tower came crashing to the ground. Just moments before, the plane had narrowly missed a tower belonging to KNOM(AM), a mile away.

The crash killed the pilot [illegible]

McBride said that the fact that the station was knocked off the air was not his first concern. "My concern was for those on the plane. In a small community like Nome, where we know most everyone, the loss of life has a great impact on all of us. While going off the air immediately changed our lives, it took several days for me to begin mourning the loss of the tower."

Race against time

KICY is a 10 kW station that [illegible] Adult Contemporary and [illegible] Contemporary music. [illegible]

FIG. 9.7 PURPOSE FOR FAA STUDIES AND HAZARD MARKING RULES (EXAMPLE FROM RCR)

The resulting cooperation between the FCC and the FAA concluded with the FCC creating new and more specific Rules in its Regulations. PART 17, Construction, Marking and Lighting of Antenna Structures, FCC Rules And Regulations (47 CFR, PART 17) . This effort also addressed the task of correlating the data bases between the FCC and the FAA for the actual locations of these and future tower sites. It appears that the tower registration program is the final phase to resolve this situation. What also may be evident is that for years there has been no fee for either of these services by either agency. With specific tower Registration Numbers and owner/location information on file, does any one want to bet that the FCC will begin to charge a fee for this "service" in the future?

Another minor "hot spot" contained on the FAA Form (FAA Form 7460-1, See Section 9.11) was the inclusion of a request for *frequency and power* information for the "proposed' facility. The question: "Include effective radiated power and assigned frequency <u>of all existing</u>, proposed or modified AM, FM, or TV broadcast stations utilizing this structure." The question created concerns that the FAA might be duplicating FCC filing information and effort.

Some charged the FAA was over stepping its bounds in requesting this type of information especially when the frequencies weren't being applied for specifically (the FCC might be assigning them at grant time). However it would not be unreasonable for the FAA to have this information as perhaps there would be some credibility for some sort of determination that possible interference could be caused to navigational systems as a result of the new facilities. So the original question remains on the FAA Form. It would be interesting to know if the FAA ever failed to approve an application based upon this frequency or power disclosure. I suppose anything is possible.

But the FAA and the FCC are not the only Federal agencies who can exercise authority over your operation when they feel they must. While operating a common carrier public access paging station in a state capitol, a Presidential visit was scheduled. The day before the arrival of the President (for a political rally as I recall) the Secret Service actually ordered our FCC licensed and Public Utility paging station "off the air", disrupting service to thousands of subscribers including many of the government officials and local law enforcement people attending the President's speech. The reason given was that our *assigned operating frequency* was "too close" to *one of their* tactical frequencies and was "locking" up their (low bid?) scanner on the "wrong" channel when we came on the air during a page. This problem is a common one when using scanners capable of receiving many frequencies over a wide frequency range. When *selectivity* is lessened in the receiver in order to receive a larger number of channels, any strong transmitter, *in-between* the desired channels scanned can *fool* the scanner into thinking its on channel.

So I remain convinced that any government agency can tell you what to do while preserving the perceived public interest even if your operation is certified, correct and official. The bottom line is that government agencies can definitely get mixed up in each other's missions and will probably continue to do so for some time.

FAA procedures are still in place and their FAA form 7460-1 must still be completed. Also new FCC forms, FCC Form 854, must also be completed so actually the work has in fact doubled and we still seem to have two government agencies requesting the same information. The only difference I can see is at least everyone is talking about the same antenna structure now even though I suspect many errors still reside in the data base with regard to actually identifying specific site locations.

9.8 Antenna Lighting and Painting (FCC)

The rules imposed by the FCC are issued pursuant to the authority of Title III of the Communications Act of 1934 which gives the FCC the authority to require painting and/or illumination of radio towers if and when, in its judgement, such towers constitute, or there is a reasonable possibility that they may constitute, a *menace* to air navigation.

The Regulation further defines antenna structures as including the radiating and/or receiving systems, supporting structures and any appurtenances (antennas) mounted thereon. What this definition does not encompass is non-radio antenna structures such as windmills, grain silos, or tall buildings. You'll find these regulations over in the FAA's Rule Book. The FCC also defines an older term: *ANTENNA FARM*, (a bit of slang?) which is a geographical location within established boundaries designated by the FCC in which a number of antenna towers with a common *impact* (poor choice of words?) upon aviation, may be grouped.

In Part 17.1, the FCC sets forth the criteria by which notification must be made to the **FAA.** Essentially any construction or alteration of more than 200 feet in height above ground level will require a new study. There are more complicated requirements for towers located near airports. These Rules deal with what is called a *glide slope* restriction. Think of it as a huge three dimensional triangle with one point resting on the end of an airport runway. As you move further from the airport, the tower height is permitted to increase along a slope within a given ratio. It is obvious that as an airplane takes off or lands the FAA does not want an antenna tower sticking up into the glide path. Any construction or alteration of greater height than this imaginary surface extending upward and outward at either end of the runway must be compliant with one of the following slope ratios: 1) 100 : 1 for a horizontal distance of 20,000 feet from the nearest point of the nearest airport runway whose length is more than 3200 feet. 2) 50: 1 for a horizontal distance of 10,000 feet for runways less than 3200 feet; and, 3) 25: 1 for a horizontal distance of 5000 feet for a heliport. These Rules are basic and many exceptions apply.

Part 17.1 also deals with such issues as towers over 1000 feet in height, exempt structures (for which an "exemption" must be approved), and structures of 20 feet or less except when it would increase the height of another antenna structure. In addition, painting and lighting requirements are specifically set forth in Subpart C of the Rules including the wattage of the lamps to be used, the intensity of the light in candelas (32.5) and the time periods for light *on* and *off* activation. Light activated devices (photocells) are even designated as being required to receive illuminance from the northern sky at a level not less than 35 foot candles (376.74 lux) for turn *on*, and a turn *off* level of 58 foot candles. The Regulations even state that *blinking* may not be more than 40 flashes per minute nor less than 12 flashes per minute with a period of darkness (in between blinks) equal to approximately one-half of the luminous period. Many of these parameters change slightly with different tower heights and so the Rules must be carefully analyzed to determine what tower lighting configuration should be used. The Rules can be applied to tower heights up to 2100 feet and above.

The Rules also provide criteria for a choice between beacons (which are red lights) or white obstruction (strobe) lighting. Variable intensity strobe systems may operate both during the daytime and evening with the intensity being dropped to night time levels of about one fifth of the day time intensity. There are significant advantages and disadvantages to using strobes over incandescent lighting systems. Maintenance , energy use and tower painting are all factors in choosing which lighting method to use and should be carefully analyzed under operational conditions. High intensity white obstruction lighting systems are primarily used for daytime illumination of towers. Generally, you do not need to paint a tower when the FCC authorizes high intensity lighting. To extend the life of the structure, however, some licensees will paint their tower anyway.

Medium intensity white lighting systems are intended for daytime, twilight or night time use. When these lights are used only during twilight and night time, the FCC may require that you **still** have to paint your tower. Generally, the FCC does not require you to paint your tower if it is 500 feet or less above the ground and the FCC has authorized medium intensity lights.

Red obstruction lighting systems are used in conjunction with tower painting. They consist of flashing red beacons and steady-burning red lights. Dual lighting systems consist of high or medium intensity lights for daytime use and red obstruction lights for nighttime use. When you are required to paint your tower you must apply the paint using alternate bands of orange and white (*peaches and cream*). You must have at least seven bands of equal width. The top and bottom bands must be *aviation orange*. The width of the bands must not be more than 100 feet nor less than 2.5 feet. You must repaint the tower whenever paint fading reduces tower visibility (pretty vague). You must also repaint the tower whenever you change the height of the tower.

Licensees who voluntarily paint and light towers when the FCC does not specifically require it must conform to the requirements specified in the Rules. Also you must notify the FAA Flight Service Station if you stop your voluntary painting and lighting the tower. Decorative marking and lighting, such as panels, call letters or decorations require written authorization from the FCC. Colored spheres (big red balls) installed on guy wires are acceptable and do not require FCC authorization.

During tower construction (Part 17.45), tower lights should also be temporarily used in some instances even if you are using an independent tower erector. Lighting requirements during construction can be critical not only because the tower is unknown to the aviation community but because if there is a problem everyone involved with it appears to be liable. A good practice is apply to the FAA & FCC before construction, notify and light up every new elevation of the tower at every construction point and notify (several times and to several locations) after each phase of construction and the overall project is complete. Document by written log what you did each day, take photographs (videotape) and record your efforts to comply with the *intent* of Part 17. These records could mean the difference between a fairly stiff fine or exoneration because of diligent behavior on your part should any mishap occur.

The only FCC citation (Notice of Violation) I ever received was for a variation in the *Aviation Orange* paint which had faded to a point where it no longer matched a **color card** used by an FCC field inspector. (Color number 12197, FS-595 can be obtained from the Specification Activity, Building 197, Room 301, Naval Weapons Plant, First and N Streets, S. E.,

Washington, D.C. 20407). The unfortunate part of the inspection process, from my perspective however, was that the shade difference didn't exactly match an unacceptable sample either. But because there was some paint pealing to a color underneath which had faded, that was the ball game. Because the time of inspection was late October, we were given ten (10) days to correct the problem. The tower was a 220 foot self supporting structure which had been selected for the installation of a medium intensity strobe system that following Spring. Painting the tower in October was our only alternative for compliance. The cost was close to four thousand dollars ($4000). Most of the paint "flaked off" the tower the following Spring anyway due to the paint being applied during freezing temperatures.

The moral of this story is try not to wait for the FCC inspector to come to you. Often you won't have time to do the job right. Because of recent collisions by aircraft (mostly private planes and helicopters) with towers, as an industry, we are about as close to *zero-tolerance* governmental Regulation as we can be. An objective of **total compliance** with tower lighting and painting regulations should be one of your highest priorities.

Perhaps one of the most important sections in the Part 17 Rules is 17.48 covering "improper operation" of your tower (or building) lighting system. The rules simply state that you must report **immediately** by telephone or *telegraph* (?), to the nearest Flight Service Station or Office of the Federal Aviation Administration, any observed or otherwise known extinguishment (outage) or improper functioning of any top steady burning or any flashing obstruction light regardless of its position on the antenna structure **within 30 minutes**. Steady burning side intermediate lights, can be corrected **as soon as possible** but notification to the FAA is not required (do it anyway!). Reporting shall set forth the condition of the light or lights, the circumstances which caused the failure, and the probable date for restoration of service. Further notification by telephone or *telegraph* (?) shall be given immediately upon resumption of normal operation of the light or lights. It is a good idea to also FAX your report to all local airports and FAA installations as well, in order to have written proof of your notification.

The FCC Rules require that you **observe the tower lights once every 24 hours** to detect outages or malfunctions. When using an automatic alarm system to detect tower light failures, you must inspect the system at least once every three (3) months. Keep a written log at the transmitter site and ask that service people complete this log every time they come on the premises at night. You can be cited even though the FCC inspector might really be impressed with your new $5000 dollar tower light monitoring system which is hooked up to your pager. He then asks to see your tower light monitor system inspection log. Now what do you do?

9.9 Antenna Site Registration (FCC)

The FCC has adopted a site registration process by which each antenna structure that requires FAA notification, including new and existing structures, **must also** be registered with the FCC by its owner. The owner is the single point of contact for resolving antenna related problems and is responsible for the maintenance of those structures requiring painting or lighting. Obtaining an antenna site registration number is a two (2) step process. The first step is to complete the FAA Study explained in Section 9.11 The second is to complete FCC Form 854 and mail it to:

Federal Communications Commission
Support Services Branch
1270 Fairfield Road
Gettysburg, PA 17326-7245

You can also obtain the form and the instructions for completion from the FCC Web site. **http://www.fcc.gov/formpage.html**

The only problem is that the form is presented as a *PDF file*. The instructions are interactive and can be down loaded. A copy of *Acrobat-tm Reader Software*, free from Adobe Systems, Inc., is what you will need. In my experienced however you'll have few problems reading the form and instructions, but unless you have access to a full version of the *Acrobat-tm* program, you might encounter some difficulty in printing. The FCC uses *Postcript, WP 5.1, Lotus 3.1 & 4.0, Form Flow, Excel 4.0 and ASCII (all-tm)* seemingly

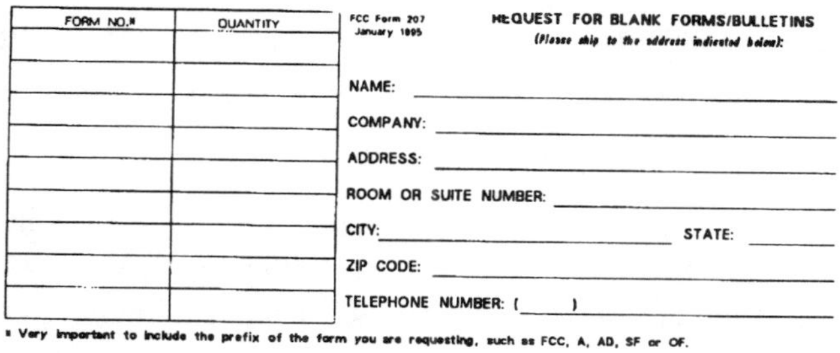
FCC Form 207
January 1995
REQUEST FOR BLANK FORMS/BULLETINS
(Please ship to the address indicated below):

FORM NO.*	QUANTITY

NAME: ______
COMPANY: ______
ADDRESS: ______
ROOM OR SUITE NUMBER: ______
CITY: ______ STATE: ______
ZIP CODE: ______
TELEPHONE NUMBER: (____) ______

* Very important to include the prefix of the form you are requesting, such as FCC, A, AD, SF or OF.

FIG.9.9 FCC FORM ORDERING BLANK

randomly for its various downloads probably because they don't want to offend any software vendors claiming favoritism. Another reason might be that they just haven't updated their forms and instructions for a while. Some of the above software might be difficult to obtain locally. Newer software versions can usually read older versions however.

You can also order copies of the form from the FCC's Form Distribution Center at 1-800-418-3676 or send your request to 2803 52nd. Ave. Hyattsville, MD 20781 (although your tower might need a second coat of paint by the time you get them from this source.)

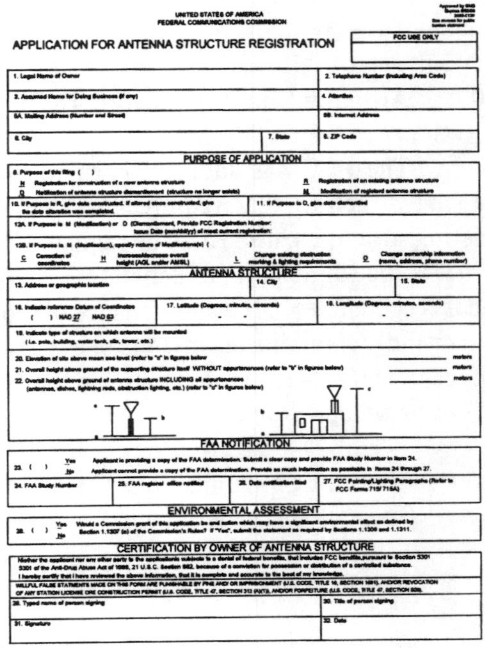

UNITED STATES OF AMERICA
FEDERAL COMMUNICATIONS COMMISSION

APPLICATION FOR ANTENNA STRUCTURE REGISTRATION

FCC USE ONLY

1. Legal Name of Owner | 2. Telephone Number (Including Area Code)
3. Assumed Name for Doing Business (if any) | 4. Attention
5A. Mailing Address (Number and Street) | 5B. Internet Address
6. City | 7. State | 8. ZIP Code

PURPOSE OF APPLICATION

9. Purpose of this filing ( )
N Registration for construction of a new antenna structure
R Registration of an existing antenna structure
D Notification of antenna structure dismantlement (structure no longer exists)
M Modification of registered antenna structure

10. If Purpose is R, give date constructed. If altered since constructed, give the date alteration was completed
11. If Purpose is D, give date dismantled
12A. If Purpose is M (Modification) or D (Dismantlement, Provide FCC Registration Number: Issue Date (mm/dd/yy) of most current registration:
12B. If Purpose is M (Modification), specify nature of Modification(s) ( )
C Correction of coordinates
H Increase/decrease overall height (AGL and/or AMSL)
L Change existing obstruction marking & lighting requirements
O Change ownership information (name, address, phone number)

ANTENNA STRUCTURE

13. Address or geographic location | 14. City | 15. State
16. Indicate reference Datum of Coordinates ( ) NAD 27 NAD 83 | 17. Latitude (Degrees, minutes, seconds) | 18. Longitude (Degrees, minutes, seconds)
19. Indicate type of structure on which antenna will be mounted (i.e. pole, building, water tank, silo, tower, etc.)
20. Elevation of site above mean sea level (refer to "a" in figures below) ______ meters
21. Overall height above ground of the supporting structure itself WITHOUT appurtenances (refer to "b" in figures below) ______ meters
22. Overall height above ground of antenna structure INCLUDING all appurtenances (antennas, dishes, lightning rods, obstruction lighting, etc.) (refer to "c" in figures below) ______ meters

FAA NOTIFICATION

23. ( ) Yes Applicant is providing a copy of the FAA determination. Submit a clear copy and provide FAA Study Number in Item 24.
No Applicant cannot provide a copy of the FAA determination. Provide as much information as possible in Items 24 through 27.
24. FAA Study Number | 25. FAA regional office notified | 26. Date notification filed | 27. FCC Painting/Lighting Paragraphs (Refer to FCC Forms 715/715A)

ENVIRONMENTAL ASSESSMENT

28. ( ) Yes No Would a Commission grant of this application be and action which may have a significant environmental effect as defined by Section 1.1307 (a) of the Commission's Rules? If "Yes", submit the statement as required by Sections 1.1308 and 1.1311.

CERTIFICATION BY OWNER OF ANTENNA STRUCTURE

Neither the applicant nor any other party to the application is subject to a denial of federal benefits, that includes FCC benefits, pursuant to Section 5301 of the Anti-Drug Abuse Act of 1988, 21 U.S.C. Section 862, because of a conviction for possession or distribution of a controlled substance.
I hereby certify that I have reviewed the above information, that it is complete and accurate to the best of my knowledge.
WILLFUL FALSE STATMENTS MADE ON THIS FORM ARE PUNISHABLE BY FINE AND/ OR IMPRISONMENT (U.S. CODE, TITLE 18, SECTION 1001), AND/OR REVOCATION OF ANY STATION LICENSE ORE CONSTRUCTION PERMIT (U.S. CODE, TITLE 47, SECTION 312 (A)(1)), AND/OR FORFEITURE (U.S. CODE, TITLE 47, SECTION 503).
29. Typed name of person signing | 30. Title of person signing
31. Signature | 32. Date

FIG. 9.9.1 FCC ANTENNA STRUCTURE REGISTRATION PROGRAM APPLICATION FCC FORM 854

FCC Form 854 is divided into six (6) sections. Section one (1) includes the legal name of the owner and the identities of persons to contact if not the same, with telephone numbers. It also asks for an *e-mail* address.

Section two (2) asks for a "purpose" of the filing, that is: N a new antenna, R a registration of an existing antenna structure, D a notification that an antenna structure no longer exists, and M a modification of a registered structure.

Section three (3) requires accurate data for the Longitude and Latitude and whether coordinates have been presented as NAD27 or NAD83 (North American Datum of 1927 or 1983) data. All heights entered on the form must be in the *Metric System,* as meters or kilometers.

To convert *feet* to *meters* multiply by 0.3048
To convert *miles* to *kilometers* multiply by 1.6093

Section four (4) allows you to attach your approved FAA Form 7460- 1 or 2 to the FCC Form 854.

Section five (5) is the standard environmental assessment question. **Hopefully, your answer will be NO!** What the FCC wants you to tell them is if your tower will kill birds, or be visible from Little Round Top at the Gettysburg Battlefield. Environmental impact can be both physical or visual. If you plan to move a lot of soil around or use high intensity strobe lights near residential neighborhoods, you could be in for some trouble. If your zoning board says its “ok” the FCC will want a copy of their approval. They also want you to include any controversial newspaper articles your tower project might have generated during the approval process. You might even have to justify your site as being in the public interest or necessity. Many times, even a remote mountain top site will be objected to merely because it can be seen by a lot of people. I had some environmentalists camping out (trespassing) around one of my towers hoping that birds blown by heavy winds and smacking into my tower could be found on the ground proving deleterious environmental effects. The fact was that some of the birds were also smacking into trees and telephone poles as well. Its pretty hard to tell at a hearing whether your tower or a nearby tree (or the plastic bag they used to store the “stunned” bird) did the poor thing in. You might want to check out Sections 1.130B and 1.1311 of Commission’s Rules to see how you should deal with any situation of this nature.

The final section six (6) of the form is your certification under oath and penalty of fine or imprisonment (U.S. CODE, Title 18, Section 1001) that you are telling the truth and that there are no willful false statements made on the form. The time to complete the form is about 30 minutes.

Telephone towers raise many concerns

What many people don't realize is that as many as three additional wireless telephone companies plan to build digital networks in the midstate.

FIG. 9.9.2 NEWSPRINT (THE PATRIOT-NEWS CO.) OF ANY CONTROVERSY SHOULD ALSO BE INCLUDED WITH YOUR FCC FORM 854 FILING

In a couple of weeks you will get an official looking green paper with your Tower Registration Number. You will be directed to furnish all your tenants a copy of this form and you will have to prove that you did it. The implication is that you must use <u>Certified Mail Return Receipt Requested</u> but an easier way is to prepare an affidavit of performance and have it notarized attaching it to the mailing list you used and the cover letter you sent. Most licensees are highly motivated to have this number anyway, so your *service* responsibility is not much of a problem. What you also might consider is a nice letter to the zoning board indicating that the FAA & FCC have both issued you a Structure Registration Number and you are forwarding a copy of your Structure Registration Certification to them for their file. Not required, but it can't hurt now can it?

FIG. 9.9.3 SITE REGISTRATION NUMBER SIGN

Finally, you must post a SIGN at your site near the structure and in plain view. These signs are approximately 12 inches by 18 inches and must meet current Part 17.4 (g) Rules concerning the marking of antenna structures. Usually it is a white background with brown borders and brown headings. The registration numbers are approximately three inches and applied as black vinyl numbers having special UV protection from the Sun so they won't fade after a few years. Check with *Artcraft Studio*, 717-545-4041 as one source for this sign.

As a postscript to the registration process, the FCC has also provided a method for electronic filing of the antenna structure data. The Web subject for this software is first the *Electronic Filing/Viewer,* then select *Antenna Structure Registration (ASR) Software* for downloading. It hardly seems worth it for only a few towers but the plus side is that you get to look at everyone else's filing. You may download the program from **http://www.fcc.gov/wtb/antstruc.html**

The software may yet be a little difficult to use and it is not known if the FCC has a working *Windows 95/98* version yet. Word has it that it also crashes easily with some hardware configurations such as a Point-to-Point Protocol (PPP) dialer. Remember you must be on-line to use this software. If you are interested in this system you can at least download the entire Public Notice released in October 2, 1996 or call the FCC Consumer Assistance staff at 1-800 322-1117 for more information. They also have a technical support group at (202) 414-1250. Additional information may be found in Fact Sheet PR 5000 #15 also available on the FCC's Web page. Oh, did I mention that you can register your antenna structure for free but that you must pay $2.30 per minute when you are reviewing or searching the data base?

The FCC has been making great strides in its data system and has recently released a Universal Licensing System (ULS). The objective of the Wireless Telecommunications Bureau in this regard is to provide an on-line data base where Applications will be filed electronically and users can receive maps or lists of licensees. The system is accessible via the FCC's site on the World Wide Web. A password, taxpayer identification number or company code is needed to access the system. There obviously are some concerns about this system including computer "hacking" changing information on Applications or frequencies in addition to some potentially serious competition with existing data providers in the commercial sector who furnish similar services. All things being equal however, it is a good first step in providing a data base system which can be used by a wide range of interested parties from engineers, attorneys and Site Acquisition Specialists.

9.10 Tower Safety Accountability

There have also been some interesting changes to Section 303 of the Communications Act of 1934 (47 U.S.C. 303) which should be noted. subparagraph (q) of this section has added the fact that not only the permitee or licensee, but in addition, *the tower owner in any case in which the owner is not the permitee or licensee*, shall maintain the painting and/or illumination of the tower as prescribed by the Commission, pursuant to this section.

In addition Section 503 (b)(5) of the Communications Act of 1934 (47 U.S.C. 503(b)(5) has also been amended to included the tower owner even if not a permitee or licensee of FCC facilities in any fines or forfeitures resulting from violations of paragraph (q) above. It means that since you own the tower the FCC can hold you accountable even if you are not a holder of an FCC license. But of course when you register your antenna structure I suppose you could be considered under FCC jurisdiction any way.

The revisions require that you notify tenants of this liability and send along copies of Part 17 of the FCC's Rules and other materials. The fines for violation of these sections can be pretty stiff and the FCC assumes that everyone including your tenants might be co-responsible. Fines as high as $20,000 per day are not unheard of but the big problem is that if insurance companies find out about your violation they may cancel your insurance. In one year alone the FCC inspected 263 broadcast towers issuing violation notices to 157 of them. The most common failure was painting or lighting problems.

Finally going back to the site rental licensing agreement, the following paragraph was recommended for inclusion by the FCC:

> Licensor (you) acknowledges that it is aware of its obligations under Section 303 of the Communications Act of 1934 (47 U.S.C. 303) to maintain the painting and illumination of the tower as prescribed by the Federal Communications Commission ("FCC"). Licensor further acknowledges that it is aware that it is subject to forfeitures assessed by the FCC for violations of such rules and requirements.

The above paragraph is your requirement but you can be sure your tenant will also want indemnification from any fines or other liabilities (doesn't everybody?) caused by your failure to comply with such requirements. Usually both parities recognize co-responsibility and agree to do what ever it takes to bring a structure back to compliance (or keep it there).

9.11 Antenna Site Determination (FAA)

On October 15th. 1992, the FAA ordered that NAD83 be used for geodetic co-ordinate information. All filings for structures which require FAA Form 7460-1 (pre-start) and FAA Form 7460-2 (begin) require a statement as to which Datum (NAD27 or NAD83) has been used in determining the structure's Latitude and Longitude. If you fail to designate this source, your application will be returned for further information causing many weeks delay. In addition the FAA from time to time releases significant changes to its rules which you must track. For example, painting of alternate orange & white bands under paragraph 33.c.1 of FAA Rules Part 77, (g) now also requires that the <u>coaxial cable, conduits and other cables attached to the face of the tower</u> also be painted. Because so many services are now being stacked on the same tower, the transmission lines (usually all black) going up the side of the tower can literally negate painting visibility. Painting transmission lines is not easy. It costs considerably more than tower members and sometimes the paint, because it is applied to a vinyl coating, peals off in a shorter period of time due to temperature extremes. If the transmission line can be pulled up inside a tower it may not have to be painted.

At a recent electronic show, I got a chuckle out of a cable vendor for suggesting that transmission line jackets be pre-colored at the factory using standard alternating color bands. I also suggested that perhaps grey vinyl be used instead of black to blend in with an unpainted tower. It is a fact that transmission lines going up the side of a self-supporting tower often presents an optical illusion to a distant observer that the tower is "leaning". More than once this illusion has resulted in a panic call to public officials that *the tower is about to fall over!*

The wording of some of these Regulations can frequently be confusing. Try this one: *the outage of one of the two lamps (required) in the uppermost red beacon or outage of any uppermost red strobe shall cause the white obstruction light system to operate in its specified "night" step intensity.* You might need help from your lighting vendor in deciphering this one.

Recently the FAA has switched the definition of *Aviation Orange* from the US Military specification which the FCC used, to the I.C.A.O. definition for the color. Other specification changes seem to be forth coming all the time such as fail safe operations, transient surge protection withstanding certain waveforms detailed in ANSI/IEEE C62.41-1991; down to the text which must be imprinted on the nameplate attached to the lighting system itself.

One sure way to insure compliance and continued understanding is to use suppliers of strobe and beacon equipment who will work with you in keeping you informed of the changes while insuring that their equipment meets all the Rules and Certifications. One such company that will supply you with excellent information in their catalog and keep you up to date is TWR Lighting, Inc. **http://twrcompanies.com**

FAA FORM 7460-1 is the primary form used in requesting an aviation study or flight hazzard determination. The form is entitled Notice of Proposed Construction or Alteration. It consists of three (3) sections.

Section one (1) identifies the site as either new or a modification of an existing site and whether the modification is temporary or permanent; also it asks when the work will begin and end. Likewise you will need to disclosure the person responsible for the filing of all future notifications as well as the frequency and ERP power level for the transmitters presently or going on the structure.

A new question asks which type of lighting system you would like to use, i.e. strobe or incandescent. If you are requesting a determination for a tall building, you better have a *side view diagram* obtainable from the building engineer or architect. Nothing fancy, just a sketch or a reduced copy of a drawing showing the proposed alteration or installation from the ground up.

Section two (2) asks for the coordinates, Latitude and Longitude, and from where they were derived. The USGS 7.5 Map is the best source, "GPS" or "Other" is not so good. The NAD determination is also critical. In addition this section asks for the nearest airport, town and directions from true north along with distance in *MILES* (not Metric). The height of the structure, ground elevation and all totals are also required in *FEET (*not Meters like the FCC).

In addition you will need a narrative description of the exact location. I always use a portion of an aeronautical map available from a local airport pilot shop, pin pointing the exact location.

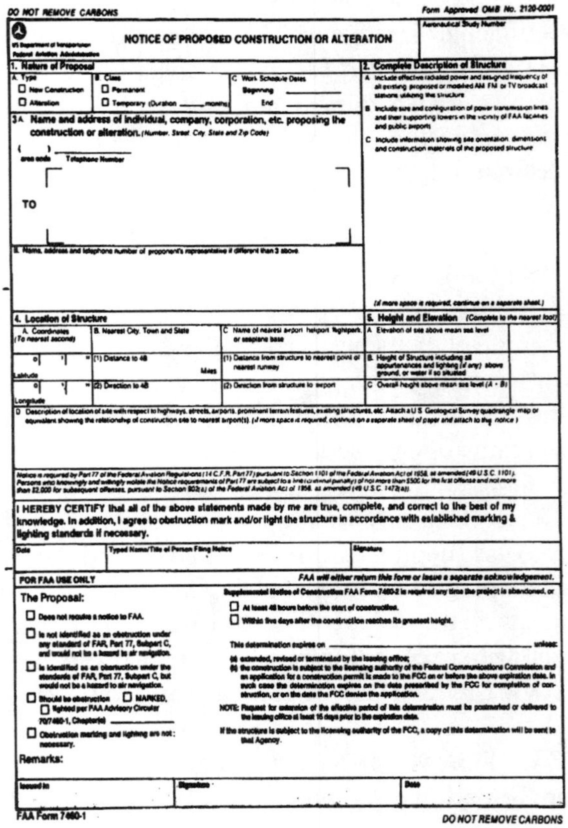
DO NOT REMOVE CARBONS — Form Approved OMB No. 2120-0001

US Department of Transportation, Federal Aviation Administration

NOTICE OF PROPOSED CONSTRUCTION OR ALTERATION

Aeronautical Study Number

1. Nature of Proposal
A. Type: ☐ New Construction ☐ Alteration
B. Class: ☐ Permanent ☐ Temporary (Duration ____ months)
C. Work Schedule Dates: Beginning ____ End ____

2. Complete Description of Structure
A. Include effective radiated power and assigned frequency of all existing, proposed or modified AM, FM or TV broadcast stations utilizing this structure
B. Include size and configuration of power transmission lines and their supporting towers in the vicinity of FAA facilities and public airports
C. Include information showing site orientation, dimensions and construction materials of the proposed structure
(If more space is required, continue on a separate sheet.)

3A. Name and address of individual, company, corporation, etc. proposing the construction or alteration. (Number, Street, City, State and Zip Code)
( ) ____ area code, Telephone Number
TO

B. Name, address and telephone number of proponent's representative if different than 3 above.

4. Location of Structure

A. Coordinates (To nearest second)	B. Nearest City, Town and State	C. Name of nearest airport, heliport, flightpark, or seaplane base
Latitude ° ' "	(1) Distance to 4B — Miles	(1) Distance from structure to nearest point of nearest runway
Longitude ° ' "	(2) Direction to 4B	(2) Direction from structure to airport

5. Height and Elevation (Complete to the nearest foot)
A. Elevation of site above mean sea level
B. Height of Structure including all appurtenances and lighting (if any) above ground, or water if so situated
C. Overall height above mean sea level (A + B)

D. Description of location of site with respect to highways, streets, airports, prominent terrain features, existing structures, etc. Attach a U.S. Geological Survey quadrangle map or equivalent showing the relationship of construction site to nearest airport(s). (If more space is required, continue on a separate sheet of paper and attach to this notice.)

Notice is required by Part 77 of the Federal Aviation Regulations (14 C.F.R. Part 77) pursuant to Section 1101 of the Federal Aviation Act of 1958, as amended (49 U.S.C. 1101). Persons who knowingly and willingly violate the Notice requirements of Part 77 are subject to a fine (criminal penalty) of not more than $500 for the first offense and not more than $2,000 for subsequent offenses, pursuant to Section 902(a) of the Federal Aviation Act of 1958, as amended (49 U.S.C. 1472(a)).

I HEREBY CERTIFY that all of the above statements made by me are true, complete, and correct to the best of my knowledge. In addition, I agree to obstruction mark and/or light the structure in accordance with established marking & lighting standards if necessary.

Date | Typed Name/Title of Person Filing Notice | Signature

FOR FAA USE ONLY — FAA will either return this form or issue a separate acknowledgement.

The Proposal:
☐ Does not require a notice to FAA.
☐ Is not identified as an obstruction under any standard of FAR, Part 77, Subpart C, and would not be a hazard to air navigation.
☐ Is identified as an obstruction under the standards of FAR, Part 77, Subpart C, but would not be a hazard to air navigation.
☐ Should be obstruction ☐ MARKED, ☐ lighted per FAA Advisory Circular 70/7460-1, Chapter(s) ____
☐ Obstruction marking and lighting are not necessary.

Supplemental Notice of Construction FAA Form 7460-2 is required any time the project is abandoned, or
☐ At least 48 hours before the start of construction.
☐ Within five days after the construction reaches its greatest height.

This determination expires on ____ unless:
(a) extended, revised or terminated by the issuing office;
(b) the construction is subject to the licensing authority of the Federal Communications Commission and an application for a construction permit is made to the FCC on or before the above expiration date. In such case the determination expires on the date prescribed by the FCC for completion of construction, or on the date the FCC denies the application.

NOTE: Request for extension of the effective period of this determination must be postmarked or delivered to the issuing office at least 15 days prior to the expiration date.

If the structure is subject to the licensing authority of the FCC, a copy of this determination will be sent to that Agency.

Remarks:

Issued in | Signature | Date

FAA Form 7460-1 — DO NOT REMOVE CARBONS

Fig. 9.11 FAA Form 7460-1 initial application for a flight hazzard determination

Section three (3) is the certification portion and signature.

Section four (4) provides the FAA response to your determination. In the upper right hand corner you will receive the all important **Aeronautical Study Number** which is used on the FCC Form 854. Local state regulatory agencies may also require notification for this study number. If your state or local government agency requires some sort of filing in conjunction with a local or municipally owned or operated airport, have your FCC Site Registration Number and your FAA Study Number before your make application if possible as things will go a lot quicker if someone has approved your project in advance.

Upon filing the form with your Regional FAA Office, (check with your local airport manager for the nearest office and the administrator in charge) you will receive either a card or an acknowledgment sheet of your filing. It will essentially repeat the information given on the form. Now is a good time to verify that the data you originally submitted is correct and input properly into the FAA system. This process is not speedy. Typically it will take 60 to 120 days for a completed study, so work ahead.

In addition the determination will expire once issued within six months if you do not file a 7460-2, Notice of Actual Construction or Alternation which is required to be submitted at least ten (10) days before the start of construction; or, within five (5) days after construction reaches its greatest height. Essentially this form fills in the dates of completion and asks you to certify that the final construction is in full accord with the original application. You should also complete this form if you do not finish construction to keep your study alive until you do. The FAA is interested in the dates that your tower reached its greatest height or if you abandon your project. You can use the form to request an extension as well but this must be approved. It is suggested you call your Regional Office with your request to extend. Failure to file these forms could bias further filings, or result in inquiries so don't forget that the entire process is one of disclosure. Should you have questions, your Regional FAA people can clear up a lot of confusion. These folks are very helpful and can get you through their process quickly. Requests for blank forms are also processed quickly.

In and around airports you may be required to provide additional information as your 7460-1 will be also routed to the Non-Rural Airport (NRA) group who will scrutinize your proposal using the tougher airport rules.

The FAA also forwards the information on your *7460* filing directly to the FCC so it is important that the two processes remain coordinated and consistent. The FAA has reported that it costs them about $120 to process a typical application and up to $2600 for more complicated ones. Accordingly, with their work load nearing 17 thousand applications per year, it is no wonder that the FAA has requested the authority from

Congress to begin charging for this service. The FAA anticipates that it will be processing about 50,000 applications each year and wanted $13 Million in fees levied for this activity.

Obviously the industry is opposing the request on the basis that payments to the FCC and local taxing authorities are already burdensome not to mention the fact that some unchallenged duplication continues between the two agencies. It is unlikely that fees will not eventually be charged for the application process in some form or another. If fees improve the processing time it might be worth it in the long run. However if the long delays in processing these applications turns out to be a leverage that the agencies want to apply to justify fees, that is another matter altogether. Demand nothing, ask a lot of questions and follow up every dialogue in writing. If you are perceived as cooperative, there is no doubt your filing will receive prompt consideration.

There will be many changes to these rules with frequent specialized procedures and form revisions enacted by these and other regulatory bodies in the near future. One excellent way to keep up to date with these changes is a close association with vendors specializing in these areas. Trade associations are also a good source of information.

Another emerging organization is *The International Engineering Consortium* which is offering a unique opportunity for learning with quick and authoritative guidelines in many technologies. Their system is almost as good as college courses on a wide range of technical telecommunication subjects. Check out their Web site at **www.iec.org** for over fifty detailed tutorials in a wide range of subjects with more on the way.

CHAPTER TEN

10.1 Profiting From Antenna Site Activities

If you've been with us to this point you already know there are a number of ways you can profit from antenna site activities. When you own and operate a site you have positioned yourself for long term and regular income. If you are willing to specialize a bit, there are other opportunities waiting in the wings which utilize your personal efforts working part-time or as a full time consultant on behalf of others.

In this chapter, we are pleased to present ten (10) scenarios which directly relate to antenna site activities. Several of these activities are obvious and the main thrust of the *Antenna Site Operating Guide.* Others are not so obvious and may involve, in addition to what you have learned about antenna sites, special abilities that you might also possess. It is a pretty good bet that you are keeping abreast of current siting events and either have some electronics or construction background in maintenance, marketing or sales; or that you have some experience in real estate, management or an interest in investment opportunities. This assumption is based upon experiences with presentations and participating in roundtable discussions on these subjects. Certain individuals always seem to come forward wanting more information. Unfortunately no material was available to offer until now. Many of the questions fielded in these presentations have served as inspiration for the preparation of this *Guide* and in some cases entire chapters. So I shouldn't think I'm too far off in assuming that some of these personalities have found their way to this chapter with similar motivations.

You might also like to consider some additional activities which are indirectly related to the antenna site business, many of which qualify as careers in their own right. The prospects for careers in *site acquisition* for example, are excellent. There are also other related jobs in the industry which can challenge both the young and the more experienced in emerging telecommunication occupations. There is room for a wide range of business and sales disciplines. Then to, if your interest is in your own business, a combination of some of these activities might also serve to assist your goal.

Finally, most of my truly rewarding work in telecommunications has taken place since my retirement. It was then that I was able to select projects of interest to me while serving a fair number of large and small communication companies. Not having to face the prospect of a boring or competitive career, project work seems to be everywhere and is exciting. You get to meet some pretty nice people and the pay is good. In most cases communication companies don't even advertise for *free lancers* because not many of them are around and when they are, negotiating a job can be a bit out of the ordinary.

If a large communication company has a small scale build-out project they won't necessarily be motivated to hire full time employees to task the job which when completed will usually be abolished. Young folks don't as a rule gravitate to jobs which transfer them all over the place only to result in layoffs when the project is completed. The use of specialized consultants solves this problem to some degree but at a premium. A *Project Coordinator* who can go anywhere, anytime on an hourly and expense basis, i.e. no retirement nor medical benefits other than disability and workmen's compensation etc., is indeed a desirable commodity. It is a great way for the semi-retired worker with communication experience to keep up with the industry while making new contacts and sharpening skills.

Consulting firms also are on the lookout for interim *Project Coordinators* too. In some cases they might be under a requirement from their client to place everyone who is associated with them under a "non-compete" or an "ideas, inventions and confidential information" (a.k.a. non-disclosure) agreement. I once started working for a site acquisition company under contract with a PCS. The work was fun the people were great but the "paperwork" didn't catch up with me until a week or two later. Buried in what amounted to be a full commission arrangement, (no salary but nevertheless fair compensation) was one of the most injurious and restrictive non-compete agreements I had ever seen. It was clear *what* they wanted me to do. But to commit to the consultant's Services Agreement for the privilege of commissioned jobs, I wouldn't be able to work for anyone else in *any* communication capacity for several years. Needless to say I didn't sign that one. Later I learned that all the PCS client wanted was a non-disclosure agreement and some sort of a fiduciary relationship between any independent contractor (me) and the consultant. Here was a case where the consultant's legal people, I feel, threw the baby (me) out with the bath water. I took another assignment and wound up working with the same people under a different arrangement and everything worked out just fine. Had I signed the thing it is unlikely I would have been permitted to publicly speak, write trade articles or even author this *Guide,* let alone work for a living.

So if you plan to work with consultants be sure to define your relationship carefully in advance. If you have local area knowledge you'll be a valuable commodity. Be in a position to negotiate a more reasonable relationship based upon an ethical devotion to your employer as well as your future value to some other company desiring your services when this contract is over. If you are a *licensed* real estate agent you might already be compliant to your employer's fiduciary relationship. After all, attorneys do this frequently and conflicts of interest are pretty easy to define in their domain. But unless you in some way actually injure your original employer it is unlikely they would challenge your non-compete status. It is an expensive process for all parties involved and is pretty difficult to defend with impunity in most states. But your talents should not be stifled nor should you be denied your future right to be able to make a living without compensation.

Here are some typical situations based upon actual cases which will allow you to evaluate how you might wish to proceed:

10.2 ONE - As A Single Tower Owner

You have uncovered a need which you have filled with 190' foot self-supporting tower and a cinder block building which you constructed on your own land. Your total cost was $62,000 of which you borrowed $50,000. You now have six (7) tenants. An FM broadcaster, two (2) paging companies, a 220Mhz. trunked radio system, a PCS and a Cellular customer. Your monthly gross rental income is $4,170 or an annualized income of *$50,040*. All tenants except the FM broadcaster and the trunked radio service pay their own electricity. Your annual costs, including taxes are $3,800 for a net income of $46,240 per year. Your plan is to pay off the tower loan in two years leaving you approximately $19,240 of net disposable income for the next two years and $46,240 yearly after that. All your leases are generally for five and ten years. Your antenna site has a market value of approximately $500,000 or 10 times "multiple". You have two more deals in the works, a business band repeater and a government agency which might add another $1000 of monthly revenue if the deals are closed. Not bad for one year of work!

10.3 TWO - As A Single Roof-top Owner

You own a four (4) story office building right along US Route 76. Your business allowed you to construct the building a few years ago for your own needs but you added several office suites as an investment in the hopes of generating some additional income in order to pay off the twenty-year mortgage. You wondered about the siting potential of your building but never realized what it might be until reading the *Guide*. Fortunately the elevator penthouse was constructed of steel girders and there are several small offices on the fourth floor near the elevator which could serve as equipment storage areas. You contacted a *site management* firm who put you under contract to represent your building with a commission of 25%. The management company procured three accounts for you, a paging receiver for a two-way paging company, a PCS site and microwave data link for a long distance carrier. Your additional net revenue is $1,725 per month for a total floor space commitment of 200 square feet. All customers *meter* their own electricity. The monthly mortgage for the building is $7,752. The net revenue from the antenna site accounts if applied to the monthly mortgage payment would offset this amount by 22.25%. However if the site revenue was used to *pay down* principal on the note an interest savings of over $223,000 would result during the life of the loan. What a great idea!

10.4 THREE - As a Five Tower *Option* Owner

You are driving along a heavily traveled twenty-five mile stretch of a rural road not too near a large city. You attempt to make a call on your cellular portable and notice the signal strength is pretty low. You call anyway but it is scratchy none the less. You know that several PCS's are developing nearby metropolitan areas and you heard that a cellular company may eventually want to improve its reliable coverage along this route (just verified by you) which it is now attempting to serve from its *boomer site* on a mountain top further away. So you resolve to try to find a few possible locations along the road.

Obtaining topographical maps showing the road, you plot the positions for several proposed towers along the way and secure options from the property owners should a tower be approved. Zoning activities are not well organized in the area so you are successful in obtaining letters of acceptance for your plan. You cost out the five (5) tower project placing them at about five mile intervals. You secure bids and run the engineering coverages at 800 and 1900 MHz. putting it all together in a composite 99% reliability multicolor map. You attach a written presentation and propose several optional lease packages which will cover your financing needs.

Site Savvy

> When your portable telephone experiences *lousy* coverage look around you, it might be a great place for an antenna site.

Remember, thus far you have done *no construction* and paid only the copy shop for the *paper package* you have developed. At this point your promotional efforts require your meeting with the players and convincing them you can have this system operational in several months. If you are successful in getting two of the six or seven players who will be developing the area, you have a plan. If you can not obtain the money to develop the system take your *options and expressions of intent* and consign them to a *tower development company* for a quick turnover deal. Should you be successful in building this package for say three (3) users on your five (5) towers, you could expect to sell the "network" for well over a million dollars in a few years.

10.5 FOUR - As a Roof-top Site Manager or a *Bulk Site* Landlord

You have several friends who own businesses, property and buildings and perhaps an apartment building in the town where you live. You offer to survey each of these properties and do a *siting assessment study* to see if any of these properties could provide site services. Several locations have provisions for equipment and antenna mounting areas on the roof and show up as being fairly usable. Others are marginal but with some owner improvement, might accommodate siting. You offer to manage these sites for a *few terms* to get things going and ask your associates if they know of any other fellow building owners who would be interested in a similar marketing analysis.. You finalize the site surveys, develop maps and profiles, post your site availability signs and put them into a site package sending them to all the current and prospective players and consultants. You *bag* several clients and implement site rentals in association with your individual roof-top property management agreements. You figure your contract commission for the buildings you now manage will total about $1250 dollars a month and that your billing services, collections and obligations to your clients will cost about $30.00 and take about one (1) hour of your time per month. You resolve to grow your building inventory and double your communication accounts each year for the next five years. At the end of your five year project period your are netting approximately $62,000 per year in commissions for your efforts.

10.6 FIVE - As a Site Deal Broker

I have a friend who has made a business of doing what consultants and communication companies can't *always* seem to do. Here is the situation. A consultant is hired by the communication company and sets up operation in the area. They hire, and bring in specialists and new prospects. The recruits usually have area real estate backgrounds or are independent contractors. After some weeks of training, the search rings are attacked and the project gets underway. My friend has taken time to develop *strong contacts* with the communication company (also likely to be consultants) and as a result gets *connected* quickly with the site acquisition key personnel.

After a time, a few of the "dogs' pop up. These are sites where building or land owners are reluctant to work with the site acquisition firm or have some personality reservations or concerns that aren't being answered. Also just about this time a few staffers with the site acquisition firm decide they are not cut out for the work leaving with some of their projects open. Prior to moving these projects to the *B-alternate site* my friend asks if he can take a crack at them before they are abandoned. After applying some plain old good salesmanship, he frequently winds up brokering the deal. The greedy site owner is usually amenable to signing *a one deal only* management representation agreement (usually because he figures he can get all the rest of the business himself) and everyone benefits.

This application really works. Here is why I believe it is successful. Some Site Acquisition Specialists are under strict guidelines and having little authority to negotiate in the field. They are also under *rent ceilings* requiring approvals. Inexperienced people are thrown into the field along with heavy site loads, expectations and performance requirements. They might encounter a recalcitrant or semi-astute business man. Time goes by with no deal. Soon a back log develops. They figure if its going to take too much time they will just *move on to the next site* and get that one done. The problem site keeps popping up on the *to-do* board and willingly assigned to my friend. The difference is that my friend is <u>not working for the site acquisition or communication company</u>. He writes his own deals. Bottom line, the site acquisition company gets the site finished, the communication company builds it and my friend snags the commission setting his now satisfied site owner up for the next deal to come along. <u>And they really do it seems.</u> I realize that this scenario might seem like a generalization but it is absolutely amazing how often a situation like this happens. Building confidences and getting things done is part of the sales and marketing discipline and not everyone is up to the task even though with the volume of work that needs to be done successes by less qualified individuals can still be achieved.

Several variations on this technique is also on my friends plate of tricks. He will routinely call a communication company project manager and ask if they have any problem sites pending or left over from a site acquisition job or just lying around waiting for some resources to pick them up again. One time he did this and got fourteen (14) search rings to work on. He didn't get them all but he landed about seven (7) of them with the difference being they were *his* deals. Not even the site acquisition company gets a piece of the action not to mention once again he now has more inventory ready for the next communication company who may want to use the sites he has obtained.

Another practice is to ask for the location (and files) for the *A sites* which were abandoned (for any number of reasons) in favor of the *B site* which was selected and constructed. Going back to the A site owner, issues are resolved. In addition, the fact that *business is being lost* is driven home providing a reality check for the *A site* owner. Often he then can be "signed" as future inventory for my friend's site portfolio.

10.7 SIX - As A Site Network Investor

Believe it or not there are lots of old tower sites and single application locations around which have been abandoned or are too small for the big boys to be interested in. In just the last few years I've reviewed hundreds of sites for sale both locally and nationally. Some of them have no services on them are in disrepair or are obsolete. As an example, microwave towers are being declared surplus by common carriers who are switching to satellite and fiber optics. Two-way shops who used to make a great living providing community repeaters have lost a lot of their business to the PCS and Cellular industry. Construction firms who used to dispatch their trucks with in-house low-band two-way radio systems have switched to ESMR and just haven't taken down their towers.

These locations may be in *grand fathered* zoning areas. In addition the facilities may be undersized and require augmentations. If the astute investor can capture a series of these facilities and network them into a *footprint offering* rather than just a single tower here or there, a marketable commodity can be presented to both the communication companies as well as tower companies who purchase these systems. In some instances small companies might *carry paper* for you, making your cash requirements significantly lower. If you have the good fortune of partnering with some investment capital, a network can be carefully optioned and acquired with you as the project developer. You might even consider using a program similar to a Real Estate Investment Trust (REIT). Here you would create stock in a closed corporation issuing it to the tower owner in return for his pledging his tower. Dividends come from site revenues. Check your local state laws for this type of operation.

One of the neatest deals that another friend of mine put together was to buy the tower site owned by a small struggling broadcast station. It seems that when the principal acquired the station he had to fiance it with a rather heavy debt load. By spinning off the antenna site for cash he could pay off the note and rent the tower (lease back) so that now his profit position was significantly improved and he could enjoy the small town status of being a successful broadcaster. Lots of opportunities here. (But be careful of AM towers unless the stations are no longer operating).

Also with the introduction of disposable circuits and microprocessor based communication, the service shop industry has changed radically in the past decade. Large communication companies have closed authorized factory service and delegated much of their servicing to different types of businesses usually by mail. Most communication companies employ their own technicians and perform wholesale equipment replacement rather than try to do field repairs with a soldering iron as once was the case. The reason is simple. The systems are too complex for field repair and too much is riding on *down time*. The loss of a single paging transmitter in a network could effect thousands of subscribers in a very small area. Thus many two-way shops are ready to close their doors or become sales agents for larger equipment suppliers. The changing nature of their business affords opportunity to acquire tower assets, materials and other supplies which have been fully depreciated and may be useful in developing your own *poorman's* antenna network. I purchased a site once and found two 180' self-supporting towers covered with weeds on the ground near the main building. These towers were new twenty years ago but had never been used. They were in fairly good shape and stored off the ground and protected. The price of a 180' tower has quadrupled over the years making the appreciation of this material asset almost as valuable as the site which was purchased. The owner didn't value these assets because they have already been depreciated and he just wanted out of the business.

CATV sites are the next big tower surplus market hitting the used tower market. Off air programing has become unreliable and satellite down stations and fibre optics are replacing many of these behemoth off-air antennas. And small systems are being consolidated and purchased by larger systems with valueless assets being spun off. Some cable companies have recognized that these old towers have some applicability, others with smaller systems could care less and don't want the responsibility of tower maintenance, FAA/FCC Regulations or to have to deal with the liability and insurance issues. The tower assets are not what is of interest but rather the number of subscribers who can be picked up in the deal. Often these towers are in *quiet* areas, i.e. not near other transmitting facilities because of their receiving functions. Then again they may not be desirable to communication carriers at all for the same reason unless they will be used as *fill-in* locations. Finally many of these towers are heavily loaded with dishes and large industrial TV receive antennas and cabling. It will be a significant expense stripping these towers to restore their loading for other clients.

10.8 SEVEN - Get A Couple of Sites at an Airport!

Several years ago I attempted to procure a receiving site at an International Airport. Most of the facilities at airports are operated by authorities or municipalities who have enough to worry about on a day to day basis without accommodating (let alone even talking to) a Site Acquisition Specialist. But there is no doubt the need for reliable airport antenna systems is growing and is very desirable to have under *your* control. In addition, any "stick" regardless of how high, anywhere on the grounds of an airport, is under the jurisdiction of the NRA (not the National Rife Association although it would seem so); but Non-Rural Airport authority of the FAA. In addition many airports are also subject to state aviation regulations or have government buildings and other regulated structures, i.e. water tanks, making the process of cutting a deal and obtaining construction and hazard marking authority most difficult. Only thing worse might be a military base, but the same principles apply.

RF considerations are also important. Even though the FAA does not exercise jurisdiction over RF matters, they are nonetheless fussy over any RF systems operating near airports and require complete disclosures, sometimes even studies on *their* forms, to prove that there will be absolutely no interference whatsoever to aeronautical air ground communication or radar. The fact that many of these systems operate in the AM mode not withstanding.

On the other hand there are thousands of people being processed through airports each hour. They make that last call or receive that last page before boarding. They can't wait till they get back on the ground to update or retrieve their messages, check their voice mail and entertain their pages etc. Waiting around airports also triggers heavy use of paging and wireless telephone services including increased on-line services through specialized carriers. Cornering the future on a suitable antenna farm location near the main terminal building and providing reliable building penetration for a myriad of low power wireless devices will put you in a very advantageous position with every carrier.

The problem is that it takes way too long to secure all the approvals even if you can rent some space in a subcontracted building nearby. Government operated systems can take months to process the simplest of requests, they usually try to get a piece of the action or set outrageous fees probably intended to frustrate your agency so they can deal directly with the communication carrier. The U.S. General Services Administration launched its National Antenna Program making roof-tops available for lease to "qualified providers" (doesn't sound like that's us) and carriers. GSA owns more than eighteen hundred (1800) buildings nationwide including some property near or on airports. Naturally these were recommendations and not regulations. It might also be necessary to deal with the specific agency housed in the building first, in addition to the GSA. Try getting into your equipment room at 11:30PM on a Saturday night. Think you'll ever get a key to the building? Here are some policy guidelines and personal comments concerning this program: development of a government lease (probably year to year); site pricing, they will study that one for at least a year; site competition, they will probably wind up using an *auction*; fee reimbursement, does the money go into the General Fund or remain with the agency using the facility such as the FAA housed in a GSA building?; site security and access rights of way, with emphasis upon *carriers fixing everything* going wrong in the building; and this one I can't help wondering about, *site request denial tracking*! Is it any wonder that communication carriers even bother with this program unless it is absolutely imperative to have the site? You can be sure there will be few twenty-five (25) year leases written on these properties. Nevertheless the GSA got a taste of the revenue possible by reporting initial leases of over $500,000 being implemented around the country. It is also a pretty safe bet that few of these leases were commissioned or are a result of site management brokers. Finally, you guessed it, they will need an office to administer *all this activity* and that means another couple of dozen government employees. Any talk of *charging fair market rates* desired by the carriers will probably give way to *percent of gross revenue*, a very distasteful concept to the industry. Seeing how states get their sales tax, and the Feds impose their excise taxes with per-subscriber fees even charged to fund the FCC, a percent of the gross revenue seems a bit much.

The U.S. Forest Service and the Bureau of Land Management along with state commissions administering public lands have been renting mountain top sites for decades. What is interesting here is that some communication firms still hold ninety-nine (99) year leases with nominal rents such as a couple of hundred dollars per year. But in recent years, these long term leases have been reduced to short periods and are offered at substantially higher rates. Work a public lease if you must but protect your future carefully.

What all this amounts too is that its pretty hard to get something happening which you can control at an airport. But if you can do it, even if it's a small local airport which someday will be huge it might be worth the effort to lock something in. I once thought of buying an old van and stuffing it with antennas and transmitters, parking it on the top level of an airport parking complex. At $7 dollars per day, that's only $220 per month. Not a bad rent if you could arrange for some electrical power.

Continuing with the project described, an increasingly pressing need for antenna sites at airport terminals is the *receive path* required in *talk-back pager* applications. After about a year, a single collinear antenna about 10' feet in length and mounted on the side of a hanger roof-top was approved. Unable to secure accessible space inside due to ownership changes, a proposal was made to use outside space for a cabinet to be mounted on a cement pad near an electrical distribution service area. The cabinet just looked like any other transformer box used for power. A water-proof cabinet can escape scrutiny so it can be enlarged to hold several rack mounted receivers, a UPS power back-up system, and a combination receiving antenna distribution amplifier and *combiner*. This way, additional two-way paging tenants can be added in *the box* without mounting another antenna structure.

Finally, some thought might be given to the installation of *triaxial* cable. These systems are also known as *radiating cables* or *leaky feeders* which provide RF coverage in enclosed or confined areas where point to point antennas are not practical. The cables have been used in tunnels, basements and in underground parking garages to provide two-way radio, paging, cellular and PCS coverage. As with all RF cables of this type degradation and power loss is of key concern. The cables are also effected by their environment and often exhibit different characteristics under differing installation conditions. Finally the cables, if installed in areas subject to fire safety regulations, must *resist flamability* and combustion problems by meeting fire retardant codes (A, RN, RNT1 and RNT); also these cables must meet NES, ASTM and IEEE standards. A good starting point for more information on this type of cable is Andrew Corporation. Ask for Bulletin 3979 for *High Performance Radiax© Cable.*

10.9 EIGHT - Brownfield Site Restoration

Here is a long shot for the strong of heart. I do not profess to know if this concept has any validity at all, but it might be worth a try if you have any experience with government agencies and the process known as *Land Recycling*. Here is the problem. Right smack in the middle of town is a site of an old ammunition factory. The soils are *loaded* with lead and other pollutants. The property has been abandoned and is now a ward of the state. The land is subject to so many state, local and federal liability issues under legislation and regulations that no one wants to touch any part of the land *let alone with a ten (10) foot pole*. The original owners have changed their names and moved out of state. Some states have initiated programs which attempt to rehabilitate the land by correcting or containing non-reversible pollutants in order to move the land use to some new long term stabilized purpose (an antenna site?). If you investigate some of these programs further there are state agencies who will work with an applicant proposal of any kind to do this, granting liability relief, engineering assistance and in some cases even financial aid and zoning waivers. The states are so glad to get rid of the problem they provide a lot of incentives to assist a site developer to contain everything and perhaps install a tower on the property. The program is usually proposal based which means you can make a written "what if" proposal without entering into any agreements. If it flies, you can move to the next step. This process is called Site-Re mediation. The authorities review the types of re-mediation required, resolve technical and liability problems, which have historically hurt any commercial real estate transaction (like zoning), and finally suggest that it saves using clean land (a.k.a. *greenfield* sites) as an alternative.

Some states work in cooperation with economic development agencies with sources of lender funds providing immunity for liability for all fiduciaries. These programs by far are not easy to sort out but they do carry the prospect of some pretty interesting alternatives which are supported by government in a lot of ways. Be sure to look around your area for any of these *brownfield* locations and make a few inquiries. One such state operating a program can be reviewed at **http://www.dep.state.pa.us**

10.10 NINE - The DTV "Kick Out"

Radio and television stations will also be effected by any *clamp down* on siting issues for the construction of new towers. The current thinking by the FCC as a result of several detailed Notices of Inquiry into the subject is that not much preemption will be taking place to brush away local government regulations. One broadcaster problem arises from the fact that many television stations will be switching over to Digital TV, a.k.a. High Definition Television or HDTV, which will require many new towers.

In general the new DTV antennas must be higher and are heavier than standard analog TV antennas. Some stations may be required to operate both the old system and the new one for a time. TV stations operating analog systems have also in the past accepted additional loading from revenue producing tenants. The fact is that in a few cases the existing tower could have been able to accommodate the new digital antenna had it not been for loading up all the other services. The TV industry must now balance the additional site revenue with the new construction costs. In some cases the cost of these new towers, some call them *super-towers*, won't even come close to the rental income generated by the site tenants. If local zoning boards are concerned over a 120' monopole they are really going to appreciate these new super towers (800 to over 1000') in their midst. The RF energy coming off one of these babies is so much greater. For example. One Million Watts ERP as opposed to the flea power generated by a PCS antenna at a mere 50 Watts or so. Under these circumstances it is hardly worth the effort considering the RFR issue at the PCS level.

The *National Association of Broadcasters* (NAB) and the *Association for Maximum Services Television* (MSTV) have filed FCC petitions asking that the FCC adopt rules preempting certain local and state restrictions on the placement, construction and modification of broadcast transmission facilities. The TV licensees may not be able to complete their upgrades as required by the industry. Of even more concern TV stations are also being stymied by local permitting authorities in making alterations and augmentations to existing structures.

"It's a bad confluence of things right now. Everybody wants tower space. Vertical real estate is like gold these days. People want space for paging, PCS and expansion of cellular phones"; says Terry Baun, of the *Society of Broadcast Engineers* (SBE) in support of the above petition.

So what does it all mean. Well for one thing if new towers are going to be constructed they are likely to be *joint ventures* by the TV stations involved. That is to say that two or three TV stations go together and build a *super tower* capable of supporting all of the DTV resources necessary. In other cases a TV station may extinguish site tenant leases who will then be scrambling for alternate sites. Finally a program of on-tower argumentation could be considered. This is a radical new technology which implements in-place tower mechanical enhancements. Henry McGinnis, a PE for *Landmark Towers*, in Ft. Worth TX said in a recent article for *Mobile Radio Technology:*

"An axiom in the tower business is that towers attract antennas and owning several towers myself, I can attest to the truth of that statement. It also has been said that necessity is the mother of invention. I know of this to be true also. One tower in our inventory was located in a popular location and reached the point of overload quickly. In fact the potential for continued loading was still growing and we determined that something innovative had to be done to bring the tower into compliance with the EIA regulations. It may even be possible to construct a tower around a tower, building up the facility's capacity to four or five times its original design.

Check with *Landmark Tower*, 1-800-438-6244 for more information.

These three situations might greatly effect site inventory. Indeed it should provide opportunity for the site developer and investor. Such *super towers* might very well be in the $1 to 5 Million Dollar range or more. This could, for example, represent an expenditure of nearly $20K to $30K in site rent (or note payments) per month for a typical fifteen year lease not to mention the required liability and casualty insurance for such a project. It is a huge financial outlay for a small local TV station. If you are into problem solving, here are some real challenges that are not likely to have clear-cut immediate solutions. An informed *deal master* might proffer some unique opportunities in this evolving situation.

There are a few national companies specializing in this process right now. But their resources might permit them to only participate in a few projects at a time. If you could even get the players to sit down to talk you certainly could proceed as a *transaction broker* in order to reach some preliminary agreements. Then invite the big boys in to see what they would be willing to do for your *consortium*. At minimum you might wind up with a commission; at best you could establish an investment package and own a piece of the deal.

10.11 TEN - Valuing a Site Business

Selling your site is one of the most difficult decisions you will have to face. But if you made the choice to sell, your next step will be to determine what you can do to increase your tower's value in the market place. The best way to accomplish this is to address weaknesses and capitalize on strengths *before* negotiations. Preparation takes some care. Assuming you understand certain potentials about your site's location an *offering notebook* should be prepared including maps, pictures, revenue projections, and expenses. You must also look at the potential for additional clients There should be some. Land values, zoning status, existing accounts, quality of leases and of course the cash flow should all be clearly stated. As a rule of thumb towers used to be valued at between 3 and 5 times annual gross income. If you had an annual gross of $50,000, a good starting place would be $250,000 in cash for the tower, property and accounts.

Other good questions to answer. Is your site reaching maturity or tower loading limits, does the site show stable growth and increasing income? Is the industry maturing with no more tenants expected for a while? Are the location and mounting opportunities limited or open? Is competition likely to happen or is the door to additional or nearby sites closing because of tighter zoning restrictions. What have you done to provide for growth or expansion. Why do tower investment companies want to pay these prices which are not as typical as in other real estate ventures or enterprises? Well for one thing they are buying your future income and that must make sense. If you have twenty(20) or twenty-five (25) year leases, they figure they will make nothing on the tower for five years and then make your income for the remaining fifteen to twenty years. If purchasing this future income doesn't make sense you won't attract any attention for your site. Secondly they are looking for additional accounts to *grow the income* to reduce the payoff time. With effective marketing, they may stand to be able to do this a lot more effectively than you could as a single tower owner.

That is why they want at least some minimum excess capacity as inventory whereby they can add tenants without much additional investment during the buy-down period. Finally they are buying a future and hopefully restricted land use situation whereby your tower might wind up being the *only ball game* in town.

There are two other considerations which you should know about as well. In the *feeding frenzy* to serve the siting needs of all these telecommunication firms, investment capital has poured into small corporations looking to go public for some really big bucks through acquisition of as many towers as possible. It seems that the best way to get into the tower business right now is to buy a tower rather than build it. Its fast, its up, its producing revenue while adding additional assets to the mix which allows additional leveraging to purchase future towers and maintain a fair credit position. Many large tower owners today merely acquire smaller tower owners and tower networks, adding each new acquisition to its nationwide portfolio of inventory. Many companies are following this line of reasoning. Investment capital does not yet seem to be a problem for either those companies who are independent or are wholly owned subsidiaries of larger companies looking to spin off with initial public offerings (IPO) or who will buy everything they can find in order to package it and then sell it to a *bigger fish up the food chain.*

Every effort should be made to a put a positive spin on these factors when offering your site for sale. You should already have some idea of the value of your site as a result of previous negotiating sessions. If there are plenty of alternatives you may have a reduced valuation. However if your tower is the only one for miles and is likely to be that way for a while you can command a greater *multiple.* But some multiples asked for by site owners have been outrageous. I saw a large tower investment company walk away from an 18 tower network offered at twelve times multiple because the towers were old, light duty, lightly loaded with inferior clientele and on land whose lease lengths were year to year and "iffy". It did however make an offer at nine times multiple for this system which was flatly rejected. Another tower investment company, in on the negotiations, took the deal at twelve. So what's going on here? Why would one firm disregard rational buy based decisions grounded upon sound fiscal and practical due diligence where another said "we'll take it"? Could it be that the 18 towers would be added to a tower portfolio and eventually wind up being combined with others, and sold to someone else for even more? You betcha! Had the second tower investment company intended to operate these facilities, the cash flow wouldn't cover the purchase price. So obviously there are other forces operating here.

Without knowing what is going on behind the scenes in the industry you would have lost some serious dollars. That is why valuation is not just based upon your property's earnings potential and can't be the only factor used when bringing it to the sales table.

Finally, what about selling your management contracts? Of course you can but they are not quite the same as hard steel inventory. A combination of both will usually fly but large tower site owners have predictably less enthusiasm over managed sites than owned *steel* sites. I often wonder why. These sites literally produce income with no plant, expenses or depreciation. They are yet another unique form of profit that can be derived from siting activities. Yet acquisition companies generally want towers. It is unlikely that managed sites would be the *only* thing they would buy. But as part of a tower deal they often acquire the whole package.

There are a great number of factors associated with the sale of a site, many beyond the scope of this *Guide*. Fortunately good accounting can come to the rescue and should be the first place you should go. Another good resource would be the previous owners of a recently acquired small tower system. These transactions are frequently reported in the trades to show momentum on the part of the company acquisitions. Large companies usually want to pay cash for your site which can create capital gains problems for you. In some instances they will suggest common stock at the current value in addition to cash. If the site company hasn't yet peaked in its *mergers & acquisition* plan it may be prudent to do a stock deal. I've seen it go both ways. But what a deal if your site is acquired, to improve the bigger company's asset position, they go public and your stock jumps twenty-points! Just hope you have held for a couple of years. So far, most site companies have not done too badly. Those that have not, just seem to get bought out.

FIG. 10.11 ROOF-TOP SITE CONTRACTS HAVE VALUE IN A FACILITIES SALE

I can't help wondering what will happen someday when one company controls 70 to 80 percent of all the antenna sites in the nation. Are we going to see an anti-trust brake up? Who knows for sure. But keeping an eye on the players and their actions will keep you pretty well informed insuring you obtain a fair market price (actual or premium) for your tower when you do decide to sell.

10.12 Related Fields and Occupations

If the preceding ten areas of income speculation were not enough to inspire you to find a niche in this fascinating and lucrative enterprise there are more areas to consider for your future in the antenna site business. These are positions needed to be filled by firms all over the country as well as independent contractor functions which you can perform by yourself, near your home, or in your own business. Some of these positions require a few skills which are not that hard to acquire. Availability is the key to success. Do they know you are available, can you do the work, are your prices reasonable. Once you get a reputation of being the contractor of choice, you are on you way to a very rewarding part time or full time career in the wireless communications field.

Let us list a few of the possibilities while briefly describing the educational requirements and the potentials for income. These are just a few and there can be many more when you consider combining the knowledge you have obtained from this *Guide* with the special skills you already possess. This synergy is almost magical in its application to the telecommunication industry and you should have little difficulty working your way into the system as a valued employee or as an independent consultant.

10.13 Site Acquisition Specialist

The goals of a professional site acquisition firm are to secure desired antenna locations ready for construction for their client communication company. The compensation to do this can vary. It is often so much per site but could be by the project, daily or even by the hour. The lines between engineering consultants preparing search rings, Site Acquisition Specialists who find actual sites and zoning activities after the site has been identified and placed under an option are beginning to blur. Many consulting firms are maturing and are now able to provide turn key services all the way to system turn on. A search of the trade publications and Web sites of some of the better known players clearly shows this evolution. But even so, the job functions are still essentially bifurcated and specialized as they always have been. However now there is better coordination between the sequential activities which may initially have had some elements of confusion or inefficiency. There are still many firms who just specialize in site acquisition or site offering services without the other activities, but most have expanded their portfolios to take a communication company from the marketing recommendation to the final facility.

Isolating just the site acquisition function for our purposes here, will enable you to locate who is working in your area, how you can present yourself and your qualifications, and perhaps sign on as a part time or full time employee or as an independent contractor.

Once hired or awarded a contract, the site acquisition company is furnished with a purpose or expectation statement and is trained in the specific protocol and systems that the communication company will be employing. It is not absolutely necessary to know the electronic theory for the systems to be used. What is important are the physical requirements and dimension for cabinets, antenna types, weight and other system parameters so that sites can be quickly evaluated to determine if they are suitable or at least augmentable to being used. Much of the planning associated with a build-out has already been accomplished prior to the arrival of the site acquisition team. That is why when certain sites can't seem to be negotiated, the entire plan or at least critical portions of it can be seriously effected. These are *key sites* and usually assigned to more seasoned analysts. In addition to critical site work, experienced analysts are also assigned the task of hiring and training additional staff as needed. After training, less experienced site acquisition personnel are relegated to less critical siting projects. In this way the major sites are acquired immediately with the peripheral or rural sites undertaken simultaneously. Rarely will a site acquisition firm bring in an entire crew from another area to do a contract job because many build-outs are going on at the same time. So only key people are brought in with a view towards picking up some temporary help in the specific location to complete the staffing. Local people can assist in many ways, area familiarity, personalities and government regulations to name a few. However a lot depends upon the deadlines imposed and the geographic area to be covered by the over all project.

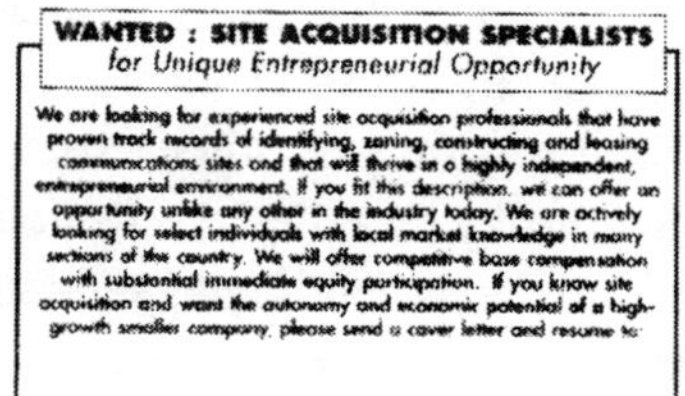

FIG. 10.13 LOOK FOR SITE ACQUISITION SPECIALIST ADS IN CURRENT TRADE JOURNALS

Then too, after a communication company is securely established in the market, a new demand arises for a *locally operated full time consultancy* to continually work on selected system build-outs and expansions. Some communication companies perform this function internally but a local consultancy might also be possible at a later date to meet these needs. In many cases, superior local site acquisition talent is either hired by the site acquisition consultant or even the communication company into full time positions. I've have seen cases where outstanding work has been recognized with a "thank you note" and a *pink slip*. However it works for you the process of making contacts and impressively performing your job goes a long way in determining what will happen after the consultant contract has concluded.

Site acquisition training programs first emphasize happiness and a good attitude toward not only the client communication company but to everyone. This emphasis is based upon the premise that conflict, personality and attitude problems have no place in this type of endeavor no matter how hard the job might be or contract is to negotiate. One must simply have to do their best regardless of how *testy* other people may become. Team work is also important as the more experienced specialist is obliged to assist the less experienced in this task oriented job.

Why even employ less experienced people? There is plenty of work that needs doing. Simply by reading this *Guide*, you probably have obtained more knowledge than the average *new hire* in an area site acquisition project. Specialization within the group promotes over all success. For example. The new employee *trudges* around the search ring, making the contacts finding the key people, spending hours researching deeds, titles and property descriptions in proximity to the desired site. A proposed contract is presented and can be meet with any number of typical responses. In some cases the site owner will just sign the contract and the deal is done. In other cases they will have it reviewed by their attorney (who may rewrite it). Renegotiations might be necessary, compromises require approvals and of course there will be a few additional rewrites along the way. Problems seem to crop up in every clause if communication attorneys do not wish to depart from their standard texts. And all this just so you can proceed with the next phase of the project, running it through zoning and building permitting process. Some consultants are likely to be considered as guest personnel in their customer's operation. The problem is that they are not employees of the host company and are often viewed as the *hired help* perhaps akin to *area temporaries* categorically being blamed for things that go wrong in any given protracted site deal. Because it is one of the first activities to take place in a PCS build-out, all eyes are upon the speed and effectiveness of actually putting the *pin into the map* demonstrating to investors and program people how close the project is to a "turn-on" and thus profitability.

We have emphasized the relationship between the systems design or RF group and their site counterparts working with the maps and search rings in the field. The locations, often referred to as *candidates,* must be researched in detail. The activity is really detailed clerical work. With a sense of understanding of the process it can be made much more effective. Further, because a site is located within the search ring it may or may not operate as predicted. To test the site, *drive tests* sometimes have to be run using small transmitting units mounted on *high-reach* or bucket trucks simulating the approximate height of the tower. Special vehicular receivers collect data as the area around the simulated site is explored. These tests are usually a condition of actual site selection and an executed contract. Because each site requires utilities, these functions must also be coordinated with telephone company and power company officials. Usually a specialized person is on the job coordinating these activities. The Site Acquisition Specialist must work with all these people as well.

Frequently site acquisition project managers will have active flow charts or Gantt charts established which guide each of the specific activities channeling the work of each interactive endeavor. These charts and work sheets vary from company to company and form the basis of a candidate *project file.* It assists the continuity of each step should a specialist be reassigned, transferred or discontinue working on the project.

Generally broad guidelines are established which allow the specialist to seek only those sites most favorable to the project, to the exclusion of others so as not to waste time. For example. It may be necessary in dense urban areas to consider only buildings that are over fourteen stories. As a rule, anything less will not be considered unless it is out in the open or positioned on a rise in topography within the city. Rural areas on the other hand might support antenna structures on roof-tops as low as four to five stories also depending upon topography and placement. *Greenfield,* or new sites, require tower and ancillary equipment, exiting sites require closer attention to equipment placement and the logistics of using the best features of the structure. Two entirely different sets of problems.

A general information form, developed a few years ago for a project transmittal in the construction of a build-to-suite program contains most of the basic *data elements* required in any site acquisition evaluation. Beginning surveys are likely to be far more detailed until the final information is "sifted down" to what is actually relevant for the given project. There can be more data for sure and the entire process after a time can be systemized into a *huge data entry* system

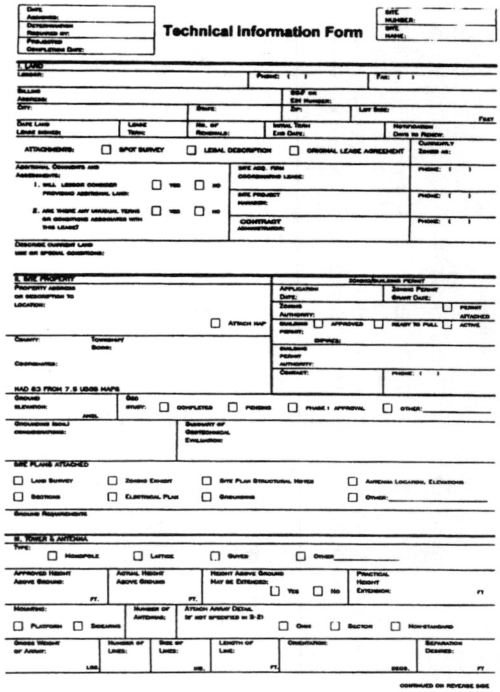

FIG. 10.13A SITE ACQUISITION SUMMARY DATA FORM

consisting of perhaps fifteen to twenty pages per site. Typically a survey form is provided by the engineering or RF group indicating their requirements. If this form passes muster, then it is on to the actual site survey inventory form. All the physical information is collected, owner's names, addresses, phone numbers, Longitude and Latitude, ground elevation, utility companies, nearby roads and other local area features. Maps, sketches and descriptions in detail of structures, locations and existing roads. Some movement into external and political factors are the next items surveyed such as the zoning area or authority. These forms can be very comprehensive and take days to research in detail. Accuracy is of the utmost importance because from this form is developed other profiles required in the process.

Some tasks performed by a typical site specialist are demanding and may require many times better than a full business day including weekends. After a time the routine is pretty straight forward as effectiveness of technique in the area improves.

Compensation can be by hour, by day, salary, or by the completed job. Typically one could expect to receive between $12 to $18 per hour, or $100 to $200 per day plus expenses as site acquisition compensation. Site acquisition work may be paid a fee for each signed site agreement which could range anywhere between $100 for a receive-only site to $2500 for a signed PCS site agreement found to be acceptable to *all* the other divisions. Now I realize that these "ball park rates" might find some challenge in the industry, but from my point of view, I wouldn't want to work for much less than the median values given. What does the site acquisition company get per site? Well that's *top secret* and a lot depends upon the way the contract was negotiated. I've heard rumors that one firm got $14,000 per completed site. Others, after rigorous competitive bidding, might get as low as $3500 per site. But look at it this way, if you are building a PCS system in an MSA of over a million population, you might need about sixty sites for a 98% penetration. If you could have them all wrapped up (not built) in 6 months for $750,000 I think you would be pretty happy. Having worked with "cut rate" site acquisition companies, I can only comment that PCS's do get what they pay for.

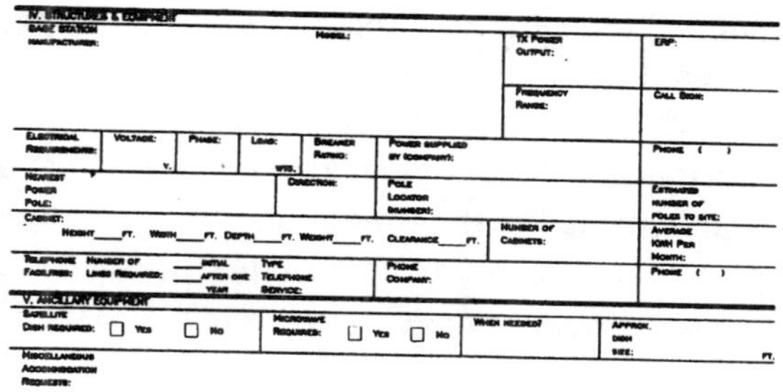
V. Ancillary Equipment
TX Power Output:
ERP:
Frequency Range:
Voltage:
Phase:
Load:
Breaker Rating:
Direction:
Pole Locator (Number):
Cabinet:
Height ___ ft. Width ___ ft. Depth ___ ft. Weight ___ ft. Clearance ___ ft.
Number of Cabinets:
Telephone Facilities: Number of Lines Required: ___ Initial ___ After One Year
Type Telephone Service:
Phone Company:
Phone ( )
Satellite Dish Required: ☐ Yes ☐ No
Microwave Required: ☐ Yes ☐ No
When Needed?
Approx. Dish Size: ft.
Miscellaneous Accommodation Requests:

FIG. 10.13B REVERSE SIDE OF SITE ACQUISITION SURVEY FORM

Salaries for full time experienced site acquisition personnel are usually in the mid $40K's to $70K's plus expenses and depending upon experience level and effectiveness. Once again *ball park salaries* but generally what is out there for quality work. A good "site-ack guy" can land about forty sites per year and with luck even more. However don't panic. Many professionals in the business know that one *really good* site can be worth more than $1 Million in revenue *per month* to a communication carrier. *So all is relative*!

10.14 Training Specialist

If you have any training or educational experience there is work for you training area Site Acquisition Specialists in the one or two week orientation program. Each company has its own requirements but I have seen several training programs and they are essentially the same. These orientation sessions cover broad subjects such as the client's specific system, procedures, reporting, negotiating skills and contracting requirements. These tasks might find you traveling all over the country for several days at a time. It can be irregular work hours too. But nevertheless interesting and directly related to many of the tasks explained in the *Guide* as supplemented by the consultant company's unique operating procedures. In some cases there are consultants who serve the consultants serving as regional trainers, researchers or seminar leaders.

The work will most likely be by the job. It allows you to interface with many of the top companies. In addition, you might like to take the information contained in the *Guide,* add it to your own experience and put your own seminar together if there is local interest. There is always public speaking potential in these subjects for civic, business, real estate and specialized groups such as builders, industrial, commercial or any other SID classification who would like to learn a lot more about antenna sites. A $10 breakfast meeting with a speaker (you) and slides covering antennas, towers, site principles, practice and potential income all expounding the basics from the *Guide*, might net you a couple of thousand per event. If you sell copies of my *Guide* while you are there, even more. *Contact me if you think you'd like to give it a try.*

10.15 Site Inspection Services

A quick review of the players in the mega-amassing of existing antenna sites nationally has demonstrated that at least a dozen or so of these mergers have taken place in recent years. Placing sites into inventory by purchasing them provides additional opportunities for site evaluation, inventory and inspection. As simple as this sounds (once again) picture a company now marketing over seventeen hundred 1700 sites from a central location or on the Internet. Companies are emphasizing ownership rather than perhaps knowing about them intimately at this point in time. Sites should be visited at least four (4) times a year, inspected for weed growth, vermin, breeches in security, lighting and a host of other preventative maintenance observations previously described in the *Guide*. In addition, sites must be visited from time to time by perspective clients, (not having keys) or for simple alarm resets, line checks etc. Take a company at random and run an area search of their owned towers from the Internet, say within a 100 mile radius of your house. Design a site inspection form and put it together with some digital pictures, an e-mail response which includes an asset report and offer it as a "Four time per year site inspection package" on a per site or package basis. If you charge $200 per tower for 4- inspections and they have sixty towers, you'll have a nice little part time business on the side. You can also be available for special site showings and "sign postings" (because they keep changing their names when new companies buy them out). There are clean-ups, trash removal or lock replacements etc.; or just a simple modification or two which service technicians sometimes do at up to $100 per hour. Can't beat this for a nice part time job. And once you obtain a reputation you can expand your work, say out to 200 hundred miles to meet additional needs. You are also free to serve other site owners as well.

10.16 Performance Technician

If you have some electronics background and wish to acquire some additional skills and equipment, you can create a highly desirable single person consultancy or take a job with a communication company in the area of transmission cable connection, repair, line sweeping and antenna testing. These functions are used when sites are first turned on, and at regular intervals to detect the degradation of transmission systems. They are also used from time to time when it is suspected that system parameters are degrading. Pre-launch testing is a detailed process of analyzing the interoperateability of all the equipment prior to its installation in the field. Once installed readings are compared and detailed to keep track of any changes which will require remedial action. Test equipment or station testers are available at substantial costs from suppliers like *Tektronix* for example but so are their training programs if you have only the theoretical background and wish to obtain some practical experience and certification. This technique is useful if your intention is to reinvent yourself or upgraded your technical skills for a more desirable avocation. Most PCS's have their own in-house network testing group. Small companies contract for these services.

FIG. 10.16 BE READY TO REPLACE TRANSMISSION LINES FOR A NICE BUSINESS

Interference studies and methods for detection take some time to learn but the process is not totally mystifying for one who would like to enter the job market. Sometimes referred to as *quality assurance* people, the process is repeated routinely. So taking this process from the construction phase to the actual optimization process can provide routine and lucrative work. But PCS and cellular testing isn't the only area of specialization. Two-way data messaging and voice paging is also a future technology with narrow band PCS (NPCS) coming along in the wings. These bidirectional communications protocols require understanding of several new technologies. Likewise the testing for quality control in these systems as opposed to PCS or analog cellular testing are a bit different. Several things to consider: a) F*unctional testing* which refers to operating the pager or NPCS device in its functional mode. b) *Sensitivity testing* for subscriber units involves the proper receiving levels and transmitting levels within a predicted coverage area. c) *Transmitter testing* includes on channel, power output, modulation characteristics and other nominal performances which must be maintained and frequently verified. *Motorola Communications* has a division called *Messaging Test Products* which provides equipment and testing instruction in these disciplines. In general these manufacturers specialize in "one-box" test equipment. Highly complex but functional pieces of equipment that test all the functions. Another manufacturer of digital radio testing is *RACAL Instruments Inc., Berkeley Varitronics Systems* **http://www.bvsystems.com** and can provide information on test units for measuring signal propagation, positioning antennas, setting power levels or validating coverages.

FIG. 10.16.1 SYSTEM OPTIMIZATION IS A LUCRATIVE CONSULTANCY

Finally, a whole new area of testing is the RF safety industry explained in detail in Chapter Nine. These studies measure Non-Ionizing Radiation (NIR) emitted from RF generators and transmitters as well as other heavy industrial equipment using RF in heating or cutting applications. One company who provides reasonably priced survey equipment and training is NARDA . Specializing in this area can provide a unique and highly desirable single person consultancy for the industry. Check in with these companies and their equipment and offerings to acquaint you with the process of system testing and the extent to which you wish to enter the activity.

10.17 Other Job Titles

In a single issue of a current trade journal the following job titles were noted:

Electronics Service Technician
Wireless Employment Opportunities - a Head Hunter
System Technicians
Manger RF Design
Network Planner
Field Service Technician
RF Engineers
Systems Engineers
Software Specialists
Site Acquisition Specialists
Zoning & Permitting Specialist
Construction Manager
Field Sales
Management Trainees Branch manager
Technical Supervisor
Bench Technicians
Microwave Installation Specialists
System Line Up tester
Base Site Installer
Switchers
RF Engineers
Performance Engineer
Implementation Engineers
Network Planners
Construction Mangers
Network Surveillance
Base Station Installer
Leasing Project Manager
Public Speaker
Marketing Manger
Retail Sales Personnel
RF Design Engineer
Telecom & Information Technology Professionals
Site Acquisition & Real Estate Specialist
Manufacturers Representative
Interconnect Engineer
A & E Manager
CADD Specialist
Radio Engineers
OSP Engineer (Fiber Optic)
Cell Technicians
Tower Riggers

Tower Painters
Tower Bulb Changers
Nationwide Agents

Most of these positions require less than two years experience. Many can be filled quickly if you are willing to travel or take on-location assignments.

10.18 Internet Web Page Designer

I won't belabor the point but this *Guide* has referred the reader to hundreds of Web sites for additional information from vendors, consultants, tower representatives and the government. One common element that immediately comes to mind when viewing only a few of these Web sites is how poorly many of them have been constructed. Further many are out of date, dysfunctional and reflect poorly the products, information, personnel and availability of information, dealers and sources of information especially for companies with whom I am familiar and who have absolutely fantastic products and services. There is a stellar opportunity for someone with Web design experience to assist these folks with far better presentations improving their image and reach ability. I needn't sell this prospective opportunity to anyone who knows what I'm saying but the niche market for work here could be a full-time telecommunication oriented Web design company. <u>Your *prospect list* is already waiting for you in Appendix B.</u>

10.19 Wireless Trading and Investing

Keeping up with the wireless industry is relatively easy compared to other more segmented industries. Many of the trade publications report literally every minor tremor, acquisition or movement of many emerging and well established companies. Not near as complex as the computer business, telecommunications, I feel, is more manageable from a trend perspective because markets are clearly assigned, more competitive and react quickly to the least bit of pressure from new entrants, services or products. In addition wireless companies are seeking good images on Wall Street to maintain their ratings and continued investor interest. Every week lists of key wireless companies are reported in *stock watch* pages and summarized in an index which has not yet emerged as a measurement in its own right. By adding less than one hundred (100) wireless companies to your stock portfolio (or practice list) you can keep track of some pretty interesting advances and investment opportunities.

For example. A trade journal reports that *360 ° Communications Co. XO* (NYSE) a cellular telephone company operating in 15 states, showed an advance for the week of 4.2%, with a 52-week low of $13 and closing at $26. This company has been acquired several times in the past and has acquired a few smaller companies itself. It has demonstrated brilliant management with adequate growth in subscribers and revenue. Now comes Alltel, *AT* (NYSE), vying for acquisition of *360 °*. In the same journal, Alltel shows a 1.8% advance for that week with a 52-week low of $29 closing at $44. The *$4.1 Billion buyout* was rumored for weeks. Plenty of time to buy-in on some stock in either company before trading might be halted. The original proposal had *360 °*share holders getting 74 cents, and a share Alltel stock. The deal is done and *360 °* is no more.

When you work numbers on some of these advance releases for various companies, the average investor could have done very well keeping track with what was going on and speculating upon the probable outcome. In fact of the one hundred or so wireless stocks of interest better than 56 had advances with less than 26 having declines. Fifteen of them had new all time highs. Even companies having a bit of trouble in the past are on the rebound according to the trades. *Moody* reported the number of issuers defaulting on publicly held corporate debt more than doubled in the past three years with technological companies representing 11% of this total. The only telecommunication company to bog down in this report was *In-Flight Phone Corp.* a provider of *Flightlink* air-to ground telecommunications. Now it is understandable that this technology may be in need of some revision by any of us who have used in-flight telephone systems. But the debt service was less than $290Million when Chapter 11 bankruptcy protection was filed in January of 1997. But think about it, the service is not going to go away and something must happen to fix it.

If you are a stay at home kind of electronic opportunist you can specialize in wireless investing and enjoy some interesting gains. There is a down-side to every story, so good knowledge of the industry doings and its players is essential.

There are also some *sleeze bags* who set up Nationwide investment schemes like the *CPA* who pleaded guilty to conspiracy and money laundering for his role in defrauding investors of an estimated $35 Million dollars. The phony wireless cable venture was telemarketed "in a manor calculated to deceive investors". The sale of these interests raised money from folks around the nation who had little or no understanding of what the services were but at least were excited enough to figure that anything in the wireless industry is "supposed to be profitable". Reports indicate that most of the money is gone. Few will be repaid. Jail terms resulting in 4 to 6 years await some of the convicted players except for one of the suspects in the case who was found dead in a Las Vegas hotel room when he failed to show up for his U.S. District Court pleading. It is sad to report that this is not the first time investment schemes in the wireless industry have surfaced though infomercial or newspaper advertisements. What is unique is that the real action with the real players is so much easier to understand and with a little research and intuition can provide a more solid investment portfolio with greater income potential. The growth potential in telecommunications is actually unbelievable. Some say the wireless market might reach $70 Billion by the middle of the next decade. It was $41 Billion in 1997 which makes the industry second only to the medial industry. If you want to position yourself where the *action is* projected, try telecommunications and in particular the antenna site industry.

10.20 Negotiating, Lobbying and Advocacy

The FCC has suggested that it promote the resolution of siting conflicts between communication companies and municipalities by using *negotiated settlements*. Stopping just short of arbitration, these activities, once developed, would then be monitored by the FCC for procedural fairness and practicality. Obviously there will be a need for skilled negotiators/arbitrators for this activity. The field is wide open on this new activity. Accordingly if you have experience in government, public service or labor negotiations, some exciting niches await you in antenna site operations as well as business.

When negotiating a contract many of the old tactics no longer apply. Deal makers rely upon honesty to make the deal work and sell. The reason is that the parties are going to be around for a long time and you have got to get off on the right foot if you want to maintain an amicable relationship with the people you have to work with. Should you be tough in your negotiations, or play the role of Mr. Niceguy? The answer to this question by people who know the negotiation business is to *do neither*. The problem is that few negotiators are well prepared and have a less than organized approach to the negotiation process.

Roger Fisher is the author of the book Getting to Yes (Penguin Books, New York City) and has been training corporate executives and labor leaders for years. The book offers methods for creating mutually beneficial agreements.

Whenever anything is to be negotiated, it is always necessary to build a *relationship* first. Real estate sales people have known this principle for years. You simply can't interact with people who have little confidence in what you do and that goes for your opponent in a negotiation. Negotiating is all about creating *value, not dividing wealth.*

Site Savvy

> Negotiating is all about creating value not dividing wealth.

It is a meeting of the minds not a competitive process. Power negotiations are sometimes created by the money involved. Perhaps a large corporation with whom you are dealing is inflexible to the point where it is left to you to be creative in order to devise another kind of advantage in the negotiation process. When you are prepared in your negotiations the people with whom you negotiate become disarmed and must acknowledge your understanding of the alternatives that they might have assumed (or were hoping) you didn't know about.

When you know with whom your are dealing, a lot more can be accomplished especially in internal negotiations . One must always have a back up plan or alternative in the event of a *deadlock* so that negotiations won't terminate abruptly. So when you are ready, listen to everything first, adapt the information, then do the deal. Take it one item at a time and keep personalities out of it. Negotiating is a fun process with much written on the subject. Learn about the fundamentals and become comfortable in the process. It is a crucial part of the site operations business and good negotiating skills can always get you what you want at prices which will make you smile.

If you have had any experience with labor negotiations or work with labor unions or operations you will have another interesting area in which to operate. In a key alliance, environmentalists teamed up with organized labor to block antenna siting and publicize alleged health risks. Why are they doing this? Well perhaps as leverage to settle more longstanding disputes related to health benefits or wage concessions with telephone owned wireless companies. The case is well known in the industry but illustrates what can happen when attention is not paid to related issues which can delay sensitive site issues. Even if the labor issues have nothing to do with siting, the sensitivity of the problem surely is a brilliant strategy on the part of the union to get management's attention. The industry warning, "look over your should for old foes with a new weapon".

Finally there are countless government issues which require an industry stated concise position and strategy. This process is called lobbying. One such issue is the Senate Commerce Committee's legislation to use Federal antenna siting fees to upgrade local and state wireless 911 systems. This is a sensitive but interesting twist. The wireless industry is now responsible for over thirty 30 million, *E911* calls per year (most of them to report auto accidents or road conditions). Some view this policy as an alternative to the FCC's providing federal preemptions to local zoning authority for the placement of towers. Could it be an "olive branch" to public service and law enforcement sectors? But predictably, the next step after this would have to be a *buck a month* on your wireless telephone bill to cover all those *E911* operators. Another issue is the imposition of the 3-percent excise tax on wireless and wireline telephone service. This tax was implemented in 1898 to fund the Spanish-American War. It was re-imposed in 1914 as a temporary tax on luxuries. But somehow it isn't clear why telephone service is still considered a luxury. Look for bill stuffers urging wireless subscribers to write their congressman in support of ending this tax. Plenty of issues to keep after. If your bent is in the advocacy area you've got your work cut out for you.

10.21 Unlocking Your Future

A couple of years ago, William C.Y. Lee, V.P. and Chief Scientist for *Air Touch Communications* speculated in an article for *Wireless World* on the future of telecommunications. The gist of his article (with apologies to Mr. Lee for being so brief):

Wireless was invented barely a century ago when Marconi successfully established an 18 mile link between the Isles of Wright and a tug boat in the English Channel. From that date to today the growth has been geometric to say the least. Are there any new technologies on the horizon which will change the basic way we communicate? Not likely. *Subspace* channels permitting faster than light telecommunications are still highly theoretical and relegated to the science fiction writer. But advances in digital and wide band technologies have revolutionized the present use of spectrum and have had some highly impressive products emerge in the last decade which promise to keep pace with the needs of a technological society becoming highly mobile. So as you can see the industry is product driven.

I can recall about 10 years ago writing a position paper for a legislative budget hearing. Operating as a communication analyst I was working under an information technology director who crossed out my use of the word "wireless" and inserted "two-way radio". What was interesting about that situation was that as an antique "wireless" collector, the word was actually one of the very first terms describing the process of radio in general (i.e. *wireless apparatus*). And now it is back with a vengeance.

So for the future we could see a *Wireless Information Superhighway System,* a single number telephone unit which follows you everywhere you go, and satellite services which provide a wealth of information technologies yet to be envisioned.

All wireless services must emanate from somewhere. This *Guide* may be the first step in assisting you in your effort to provide for some of these services through the building, marketing and operating of an antenna site for profit. Good luck!

APPENDIX A COMMUNICATION COMPANIES

These company listings represent only the major players in national telecommunications build-outs. Check local directories, trade journals and industry compilations for hundreds more.

Nextel Communications Inc.
1505 Farm Credit Drive
McLean, VA 22102
(703) 384-3000

21st Century Wireless Group Inc.
406 Gateway Blvd.
Burnsville, MN 55337
(612) 890-8800

RSC Communications Inc.
3855 S. Northern
Independence, MO 64052
(816) 313-8255

Frontier Cellular
161 Chestnut Street
Rochester, NY 14604
(908) 919-1000

Bell Atlantic Nynex Mobile
180 Washington Valley Rd.
Bedminster, NJ 07921
(908) 306-7000

BellSouth Cellular Corp.
1100 Peachtree St.
Atlanta, GA 30309
(404) 249-5000

GTE Mobilnet
245 Perimeter Center Parkway
Atlanta, GA 30346
(770) 391-8000

Ameritech Cellular Services
2000 W. Ameritech Center Drive
Hoffman Estates, IL 60195
(847) 706-7600

Vanguard Cellular Systems
2002 Pisgah Church Road
Greensboro, NC 27455
(910) 282-3690

SNET Mobility Inc.
500 Enterprise Drive
Rocky Hill, CN 06067
(860) 513-7600

Alltel Communications
One Allied Drive
Little Rock, AR 72202
(501) 661-8000

AirTouch Communications Inc.
One California St.
San Francisco, CA 94111
(415) 658-2000

Century Cellunet Comm. Inc.
100 Century Park Drive
Monroe, LA 71203
(318) 388-9500

Comcast Cellular Comm. Inc.
480 E. Swedwesford Road
Wayne, PA 19087
(610) 995-5000

Western Wireless Corp.
330 12th. Ave. N.E. #200
Bellevue, WA 98005
(206) 313 5200

United States Cellular Corp.
8410 W. Bryn Mawr.
Chicago, IL. 60631
(773) 399-8900

CommNet Cellular Inc.
8350 E. Crescent Pkwy.
Englewood, CO 80111
(303)694-3234

U S West New Vector Group Inc.
3350 161st Ave S.E.
Bellevue, WA 98008
(205) 747-4900

Price Comm. Wireless Inc.
12800 University Dr.
Fort Myers, FL 33907
(941) 433-4350

SBC Wireless Inc.
17330 Preston Road
Dallas, Texas 75252
(972) 733-2000

Pri Cellular Corp.
711 Westchester Ave.
White Plains, NY 10604
(914) 422-0800

Radiofone Inc.
3131 N I-10 Service Rd. E
Metairie, LA 70002
(504) 837-8330

Triad Cellular Corp.
2420 Sand Hill Road
Menlo Park, CA 94025
(415) 324-6885

DCC PCS Inc.
13439 N. Broadway Extension
Oklahoma City OK 73114
(404) 391-8400

Mercury Mobility L.L.C
One Lakeshore Drive
Lake Charles, LA 70629
(318) 436-9000
Fortunet Comm. L.P.
350 Stuyvesant Ave.
Rye, NY 10580
(914) 967-1606
Telecorp Holding Corp. Inc.
1110 N. Glebe Road
Arlington, VA 22201
(703) 351-5100
BellSouth Mobility PCS
3350 Peachtree Road NE
Atlanta, GA 30326
(404) 841-2000
Genberal Wirless Inc.
8144 Walnut Hill Lane
Dallas, TX 75231
(214) 265-2550
Cox Communications Inc.
18200 Von Karmen
Irvine, CA 92612
(714) 623-5000
Aer Force Communications LP
350 Stuyvesant Ave.
Rye, NY 10580
(914) 921-6300
US West Communications Inc
1999 Broadway 10th Floor
Denver, CO 80202
(303) 294-1612
Atlantic Cellular
15 Westminister St.
Providence, RI 02903
(401) 458-1900
Chase Telecommunications Inc.
64420 Richmond Ave.
Houston TX 77057
(713) 782-3332
Magnacom Wireless Inc.
4001 main Street
Vancouver, WA 98663
(360) 694-9438
McLeod USA Inc.
6400 C St. SW
Cedar Rapids, IA 52404
(319) 298-7000
Powertel Inc.
1233 O.G. Skinner
West Point, GA 31833
(706) 645-2000

Aerial Communications Inc.
8410 W. Bryn Mawr Ave.
Chicago, IL 60631
(773) 399-4200
Pacific Bell Mobile Services
4420 Rosewood Drive Bldg.2
Pleasanton, CA 94588
(510) 227-3300
21st Century Telesis Inc.
650 Town Center Drive
Costa Mesa, CA 92626
(714) 752-2178
Urban Communications PCS LP
505 8th. Ave.
New York, NY 10018
(212) 714-1000
Ameritech Cellular Services
2000 W. Ameritech Center Drive
Hoffman Estates, IL 60195
(847) 706-7600
Century personal Access Network
100 Century Park Drive
Monroe, LA 71203
(318) 388-9000
Rivgam Communications LLC
8 Sound Shore Drive
Greenwich, CT 06830
(203) 629-3333
Pocket Communications Inc.
2550 M Street NW Suite 200
Washington, DC 20037
(202) 629-4300
Northcoast Operating Co. Inc.
6800 Jericho Turnpike
Syosset, NY 11791
(516) 393-5806
Western Wireless Corp.
2001 NW Sammamish Road
Issaquah, WA 98027
(206) 313-7720
PrimeCo Personal Comm. LP
6 Campus Circle
Westlake, TX 76262
(206) 313-7720
Phillie Co LP
4717 Grand Ave.
Kansas City, MO 64112
(816) 559-1000
Devon Mobile Comm. LP
Main & Water Street
Coudersport, PA 16915
(814) 274-9830

Comcast PCS Communications Inc.
480 E Swedesford Road
Wayne, PA 19087
(610) 955-5000
GTE Wireless Inc.
600 Embassy Row
Atlanta, GA 30328
(770) 379-2800
Mercury PCS LLC
1410 Livingston Lane
Jackson, MS 39213
(601) 362-2200
Cook Inlet Western Wireless
2001 NW Sammamish Road
Issaquah, WA 98027
BellSouth Mobility Inc.
1100 Peachtree Street NE
Atlanta, GA 31833
(706) 249-5000
NextWave Telecom Inc.
6256 Grenwich Dr.
San Diego, CA 92122
(619) 559-2828
Sprint Spectrum LP
4717 Grand Ave.
Kansas City, MO 64112
(816) 559-6790
AT&T Wireless Services Inc.
5000 Carillon Point
Kirkland, WA 98033
(206) 827-4500
Omnipoint Communications Services
16 Wing Drive
Cedar Knolls, NJ 07927
(201) 257-2400
Pittencrieff Comm. Inc.
1 Village Drive
Abeline, TX 79606
(915) 690-5800
Industrial Communications & Electronics Inc.
100 Marion Drive
Kingston, MA 02364
(617) 585-9100
Advanced Electronics
18220 S. Broadway
Gardena, CA 90248
(310) 532-3211
Skitronics Inc.
4716 Fayetteville Road
Raleigh, NC 27603
(919) 662-3220
SMR Direct
1600 Wynkoop Street
Denver, CO 80202
(303) 571-5050
Porta-Phone
1306 Thomasville Road
Tallahasee, FL 32303
(904) 385-6400
Wireless Plus Inc.
26102 Eden Landing Road
Hayward, CA 94545
(510) 782-5000
Southern Communications Inc.
64 Perimeter Center East
Atlanta, GA 30346
(770) 821-4800
Fisher Communications Inc.
14530 S. Commercial St.
Blythe, CA 92225
(619) 922-4150
Sierra Communications
3167 Fitzgerald Road
Rancho Cordova, CA 95742
(916) 631-0160
Applied Technology Group Inc.
4440 Easton Drive
Bakersfield, CA 93309
(805) 322-8650
Lagorio Communications Inc.
2771 E. French Camp Road
Manteca, CA 95336
(209) 982-1993
Constant Communication
705 N. Lamar
Austin, TX 78703
(512) 495-1333
Advanced Communications Inc.
593 Overmyer Road
Sparks, NV 89431
(702) 353-3600
Cellular Design Corp.
P.O. Box 591
Deer park, NY 11729
(516) 242-5900
Triangle Communications Inc.
940 West Main Street
New Holland, PA 17557
(717) 656-2211
Paging Network Inc.
4965 Preston Park Blvd.
Plano, TX 75093
(972) 985-4100

Arch Communications Group Inc.
1800 West park Drive
Westborough, MA 01581
(508) 870-6700

Paging Partners Corp.
4249 Rte. 9 North, Bldg. 2
Freehold, NJ 07728
(908) 409-7088

Southern Net Association of Paging
1604 Chickasaw Street
Greenwood, MS 38930
(601) 453-0662

Bell Atlantic Paging Inc.
1719A Route 10, Suite 300
Parsippany, NJ 07054
(973) 560-8000

Network Services LLC
525 S. Douglas Street
El Sequndo, CA 90245
(310) 615-6500

Priority Communications Inc.
1499 W. Palmetto Park Road
Boca Raton, FL 33486
(561) 750-8899

Teletouch Communications Inc.
110 N. College
Tyler, TX 75702
(903) 595-8800

MobileMedia Communications Inc.
65 Challenger Road
Ridgefield Park, NJ 07660
(201) 440-8400

AirTouch Paging
12221 Merit Drive
Dallas, TX 72521
(972) 860-3219

Metrocall Inc.
6677 Richmond Highway
Alexandria, VA
(703) 660-6677

PageMart Wireless Inc.
6688 N. Central Expressway
Dallas, TX 75206
(214) 750-5809

TSR Wireless Inc.
400 Kelby Street
Fort Lee, NJ 07024
(201) 947-5300

ProNet Inc.
6340 LBJ Freeway
Dallas, TX 75240
(972) 687-2000

Preferred Networks Inc.
850 Center Way
Norcross, GA 30071
(770) 582-3500

SourceOne Wireless Inc.
1040 S. Milwaukee Ave.
Wheeling, IL 60090
(847) 403-7000

American Paging Inc.
1300 Godward Street
Minneapolis, MN 55413
(612) 623-3100

Mobile Telecomm. Tech. Inc.
200 South Lamar
Jackson, MS 39201
(601) 623-3100

AT&T Wireless Services Inc.
5000 Carilon Point
Kirkland, WA 98033
(425) 828-8476

American Moble Satellite Corp.
10802 Parkridge Blvd.
Reston, VA 22091
(703) 758-6000

Celsat America Inc.
3460 Torrance Blvd
Torrance, CA 90503
(310) 316-1339

Globalstar L.P.
3200 Zanker Road
San Jose, CA 95164
(408) 473-5550

Iridum L.C.
1575 Eye St. NW
Washington, DC 20005
(202) 408-3800

Odyssey Telecomm. Int. Inc.
One Space Park
Redondo Beach, CA 90278
(310) 814-8888

Chadmoore Wireless Group Inc.
2875 East Patrick Lane
Suite G
Las Vegas, NV 89120
702-740-5633

APPENDIX B WEB SITES

As always Web addresses and names change regularity. Should you discover that a given Web address is no longer valid, one of the best sources for updated information are listings presented in the *Wireless Week* Web page at **www.wirelessweek.com** also *RCR* at **http://rcrnews.com** . Downloading these files into a word processor can also activate each address for *hyper-link* selection so that double-clicking on the address will take you directly to the site through your word processor and ISP. The sources for these Web sites include company catalogues, advertisements, correspondence, business cards, *WirelessWeek*, *RCR* and IWCE conference materials. Every effort was made to include accurate addresses but the author makes no warrantees as to the operation of the site or the material included herein. For a complete listing of the companies serving your area consult the above two web sites for data relative to your market.

Communications Companies - (use for you solicitations)

21st Century Communications.	http://www.pagers.tetranet.net.
21st Century Wireless Group Inc.	http://www.21wireless.com.
A+ Network Inc.	http://www.aplsnet.com.
Aerial Communications Inc.	http://www.aerial1.com.
Airadigm Communications Inc.	http://www.einsteinpcs.com
AirCell Inc.	http://www.aircell.com.
Airmail Communications.	http://www.airmailcomm.com
Airtouch Comm. Inc.	http://www.airtouch.com
AirTouch Satellite Services.	http://www.airtouch.com/globalstar
Aliant Communications Inc.	http://www.aliant.com
Alltel Corp.	http://www.alltel.com
American Mobile Satellite Corp.	http://www.skycell.com
American Paging.	http://www.americanpaging.com
Ameritech Corp.	http://www.ameritech.com
Angellcom	http://www.angellcomcom
Arch Communications.	http://www.arch.com
Ardis Co.	http://www.ardis.com
AT&T Corp.	http://www.att.com
AT&T Wireless Services Inc.	http://www.attws.com
AT&T Wireless Services Inc.	http://www.airdata.com
Bay Star Satellite Comm.	http://www.baystar.com
Bell Atlantic Mobile.	http://www.bam.com
Bell Mobility Inc.	http://www.bellmobility.com
BellSouth Cellular Corp.	http://www.com/bscc
BellSouth Corp.	http://www.bellsouth.com
BellSouth Mobility.	http://www.bellsouthdcs.com
Bluegrass Cellular Inc.	http://www.bluecell.com
British Telecom	http://www.bt.com
BTI.	http://www.btitele.com
Cable and Wireless PLC.	http://www.cwplc.com
The Carlyle Group	http://www.thecarlylegroup.com
Cantel.	http://www.cantel.com
Cellular One Group.	http://www.cellularone.com
Cellular One of AZ, NM &TX.	http://www.cell1.com
Cellular One of Greater MN.	http://www.starpoint.net/~cellular.
Cellular One of San Francisco.	http://www.cellone-sf.com
Centennial Cellular Corp.	http://www.centennial.com
CFW Comm. Co.	http://www.cfw.com
Clearnet Comm. Inc.	http://www.clearnet.com
Comcast Cellular Comm. Inc.	http://www.comcastcellular.com
CommNet Cellular Inc.	http://www.cels.com
Comsat Corp.	http://www.comsat.com
Contact Paging.	http://www.contact-paging.com
Cue Network Corp.	http://www.cue.net
Datapage Inc.	http://www.datapage929.com
Debit Cellular Network.	http://www.dcnnet.com

Digiph.	http://www.digiph.com
Direct Comm Inc.	http://www.dcsat.com
El Paso Paging.	http://www.elpasopaging.com
Flagship Wireless.	http://www.federalnetwork.com
Frontier Corp.	http://www.frontiercorp.com
General Wireless Inc.	http://www.general-wireless.com
Geotek Comm. Inc.	http://www.geotek.com
GeoTrans Wireless.	http://www.geotranswireless.com
Globalstar Tele.	http://www.globalstar.com
GTE Corp.	http://www.gte.com
GTE Wireless.	http://www..gte.com
Heartland Comm.	http://www.heartcomcom
Hellyer Comm. Inc.	http://www.hellyer.com
Horizon Personal Comm. Inc.	http://www.horizonpcs.com
ICG Satellite Services.	http://www.icgws.com
Intellicom	http://www.intellicomm220.com
InterCel.	http://www.icel.com
Int. Maritime Sat. Org.	http://www.inmarsat.org
Int. Tele. Sat. Org.	http://www.intelsat.com
Iridium LLC.	http://www.iridium.com
Leo One USA Corp.	http://www.leoone.com
LocusOne Comm.	http://www.locusone.net
Maine Wireless.	http://www.mainewireless.com
MCI Comm. Corp.	http://www.mci.com
Media Page.	http://www.datadepot.com
Metricom Inc.	http://www.metricomcom
Metrocall Inc.	http://www.metrocall.com
Microspace Comm. Corp.	http://www.microspace.com
Midwest Wireless Comm.	http://www.mwcom.com
Mobilecomm Inc.	http://www.mobilecomm.com
MobileMedia Comm. Inc.	http://www.mobilemedia.com
MobiLink.	http://www.mobilink.com
Mobilphone de Panama.	http://www.mobilphone.com
National Dispatch Center Inc.	http://www.ndcwireless.com
Nextel Comm. Inc.	http://www.nextel.com
NextWave Telecom Inc.	http://www.nextwavetel.com
North Sight Comm. Inc.	http://www.northsight.com
NPI Wireless.	http://www.noverr.com
Nynex Corp.	http://www.nynex.com
Omnipoint Corp.	http://www.omnipoint.com
Pacific Bell Corp.	http://www.pacbell.com
Pacific Telesis Group.	http://www.pactel.com
PageMart Canada.	http://www.pagemart.ca.
Pagemart Inc.	http://www.pagemart.com
Paging Network Inc.	http://www.pagenet.com
Personal Comm. Inc.	http://www.pagepci.com
PetroCom	http://www.petrocomcom
Pittencrieff Comm. Inc.	http://www.pitc.com
Pocket Communications.	http://www.pocketcomm.com
Pocono Mountain Comm.	http://www.pocmtn.com
Police Emergency Radio Svs. Inc.	http://www.intellicomm220.com
Positive Comm. Inc.	http://www.positive.com
Powertel.	http://www.powertel.com
Preferred Networks Inc.	http://www.pni.net.
PrimeCo Personal Comm. LP.	http://www.primeco.com
ProNet Comm.	http://www.pronetcomm.net.
RAM Mobile Data.	http://www.ram.com

RAM Technologies Inc.	http://www.ramlink.net.
Rogers Comm.	http://www.rogers.com
Rural Cellular Corp.	http://www.rccwireless.com
SatCom Systems Inc.	http://www.satsrv.com
SBC Communications Inc.	http://www.sbc.com
Secure Cellular Inc.	http://www.pagers.com/secur.html.
Shaw Communications.	http://www.shaw.ca.
SK Telecom.	http://www.sktelecomcom
Skysite Comm. Corp.	http://www.skysite.com
SkyTel Corp.	http://www.skytel.com
SMR Direct.	http://www.smrdirect.com
SourceOne Wireless.	http://www.sourceonewireless.com
Southern Comm. Svs Inc.	http://www.solinc.com
Southern New Eng. Tele. Corp.	http://www.snet.com
Southwestern Bell Tele.	http://www.swbell.com
Southwestern Bell Wireless.	http://www.swbellwireless.com
SpaceCom Systems Inc.	http://www.spacecomcom
Sprint Corp.	http://www.sprint.com
Sprint PCS.	http://www.sprintpcs.com
Sprint Spectrum/Amer. P/C	http://www.sprintspectrum-apc.com
Star Paging Inc.	http://www.usstar.com
StarCellular.	http://www.starcellular.com
Teledesic Corp.	http://www.teledesic.com
Teletouch Comm. Inc.	http://www.teletouch.com
Telstra Corp.	http://www.telstra.com
TSR Paging Inc.	http://www.beep.com
United States Cellular.	http://www.uscc.com
US West Inc.	http://www.uswest.com
USA Wireless.	http://www.usawireless.com
ValuePage Inc.	http://www.valuepageinc.com
Vanguard Cellular Sys. Inc.	http://www.cellonepa.com
Vodafone Group.	http://www.vodafone.co.uk.
WebCel Comm. Inc.	http://www.webcel.com
Western Wireless Corp.	http://www.voicestream.com
WinStar Comm. Inc.	http://www.winstar.com
Wireless 101.	http://www.wireless101.com
Wireless One Net.	http://www.wirelessonenet.com
Wynd Comm. Corp.	http://www.wynd.com

Additional Government, Manufacturers, Consultants, Suppliers Site Management and Communications Equipment Vendors:

3COM.	http://www.3com.com
Absolute Time Corp.	http://www.absolutetime.com
AccessLine.	http://www.accessline.com
Accugraph Corp.	http://www.accugraph.com
ACE Antenna.	http://www.aceantenna.com
ACS Wireless Inc.	http://www.acs.com
ACT Com	http://www.actComm..com
ADC Metrtica.	http://www.metrica.com
ADC Telecom Inc.	http://www.adc.com
ADC Wireless Systems.	http://www.wireless.adc.com
ADS Power Resource.	http://www.adspower.com
Adtran Inc.	http://www.adtran.com
Advanced Battery Sys.	http://home.navisoft.com/periphex
Advanced Cellular Comm. Corp.	http://www.accc.com
Advanced Fox Antenna.	http://www.advfoxcell.com

Advanced Radio Systems Corp.	http://www.arsradio.com
Advanced Radio Telecom Corp.	http://www.artelecomcom
Advanced Signal.	http://www.advsig.com
Advanced TechCom Inc.	http://www.atiradio.com
Advanced Videotech Corp.	http://www.advfoxcell.com
Advancetec Industries Inc.	http://www.advancetec.com
Advantage Wireless.	http://www.adv-wireless.com
Aerospatiale.	http://www.aerospatiale.fr
Aerotron-Repco Systems Inc.	http://www.aerotron-repco.com
Aether Technologies.	http://www.aethertech.com
Aether Technologies.	http://www.aethertech.com
AF CommSupply.	http://www.afcommsupply.com
Affordable Message Center.	http://www.amcalpha.com
AG Comm. Systems.	http://www.agcs.com
Aims Technologies Inc.	http://www.aimstech.com
Air Technology.	http://www.pureair.com
Aircom International.	http://www.aircomco.uk
AirLink Comm. Inc.	http://www.airlink.com
AirMedia Inc.	http://www.airmedia.com
AirMedia Notify! Group.	http://www.exmachina.com
AirNet Comm. Corp.	http://www.airnetcomcom
AirNote Wireless Email.	http://www.airnote.net.
Aironet Wireless Comm. Inc.	http://www.aironet.com
Airtech Cellular Ltd.	http://www.airtechwireless.com
Alan Dick & Co. USA Inc.	http://www.alandick.com
ALcatel.	http://www.ans.alcatel.com/products/mw
Aldiscon Inc.	http://www.aldiscon.com
Alexander Batteries.	http://www.alexmfg.com
Allen Telecom Group.	http://www.allentele.com
Allgon Systems.	http://www.allgon.com
Allied Tower Corp.	http://www.alliedtower.com
Alltel Corp.	http://www.alltel.com
Alpha Industries.	http://www.tiac.net/users/ppiro
AltoScience Inc.	http://winston7.home.mindspring.com
Amdocs Inc.	http://www.amdocs.com
American High Mast Systems.	http://www.ahms.com
American Line Builders Inc.	http://www.albinc.com
American Manage. Systems Inc.	http://www.amsinc.com
American Reliance Inc.	http://www.amrel.com
American Skynet.	http://flint.org/propage.
American Tower Systems Inc.	http://www.amertowersys.com
AML Comm..	http://www.amlj.com
Amphenol Products Corp.	http://www.amphenol-aipc.com
Amplifier Research.	http://www.ar-amps.com
Amtelco.	http://www.amtelcomcom
Amtronix Test Equipment.	http://www.madbbs.com/amtronix
Analog Devices.	http://www.analog.com
Anchor Graphics.	http://www.anchorgraphics.com
AnchorGuard.	http://www.anchorguard.com
Andrew Corp.	http://www.andrew.com
Angellcom.	http://www.angellcomcom
ANI Site Development.	http://www.anisite.com
Anritsu Wiltron.	http://www.anritsuwiltron.com
An-ser Services.	http://www.anser.com
Answermart/American Beepers.	http://www.answermart.com
Antenex Inc.	http://www.antenex.com
Antenna Sites Inc.	http://www.antenna-sites.com

Antenna World Inc.	http://www.antennaworld.com
Apac TeleComm..	http://www.apac-tele.com
A.P.E.	http://www.apecorp.com
Apex Site Management	http://www.apexsite.com
Applied Energy Concepts Inc.	http://www.aeconcepts.com
Applied Innovation Inc.	http://www.aiinet.com
Argus Tech. Ltd.	http://www.argus.ca
Arianespace.	http://www.arianespace.com
Aris Inc.	http://www.arisinc.com
Arjay Telecom	http://www.arjay.net
Arjay Wireless.	http://www.arjaywireless.com
ArrayComm Inc.	http://www.arraycomm.com
Ashtech Inc.	http://www.ashtech.com
Astrolab Inc.	http://www.astrolab.com
Astron Antenna Corp.	http://www.astronantennas.com
Astronet.	http://www.astronetcorp.com
ATC.	http://www.atct.com
Atlantic Comm. Inc.	http://www.aciaccess.com
Atlantic Scientific Corp.	http://www.iu.net/atlsci
Atlas Systems.	http://www.intraxel.com
Atlas/Soundolier.	http://www.atlas-soundolier.com
ATX Technologies Inc.	http://www.track.com
Audiovox Corp.	http://www.audiovox.com
Auto Club Cellular.	http://autoclub1.com
Automotove Technologies Inc.	http://www.thecarphonestore.com
A.V. Lauttamus Com	http://www.lauttamus.com
Avcom of Virginia.	http://www.avcomofva.com
AvData Systems Inc.	http://www.avdata.com
AVS Wireless Services.	http://www.thevision.net/avs
Avtec.	http://www.avtecinc.com
Axiom Inc.	http://www.axiomav.com
Aydin Tele. Microwave Products.	http://www.aydin.com
B&C Comm..	http://www.bc-comcom
Badger Technologies Inc.	http://www.badger.com
Badger Towers.	http://www.badgerland.com/towers.htm
Badgerland Comm. Inc.	http://www.badgerland.com
Baja Sun.	http://www.suntech.com
Ball Wireless Comm. Group.	http://www.ball.com
Ballantine Labs Inc.	http://www.ballantinelabs.com
BAM Computer Solutions Inc.	http://www.bamcs.com
Barcodes West Inc.	http://www.bcw.com
Barnett Electronics Inc.	http://www.barnettelec.com
Battery Network.	http://www.battnet.com
BatteryPRO Systems Inc.	http://www.batterypro.com
BCS Wireless.	http://www.bcswireless.com
BearCom	http://www.bearcomcom
BECO Rooftop Leasing.	http://www.beconet.com
Belkin Components.	http://www.belkin.com
The Bentley Co.	http://www.bentleycompany.com
Berkeley Varitronics Sys. Inc.	http://www.bvsystems.com
Berliner Com Inc.	http://www.bcisites.com
Best Power.	http://www.bestpower.com
BIA Telecom Inc.`	http://www.biacompanies.com
Biby Engineering PC.	http://www.biby.com
Big Bear 2-Way Comm..	http://www.bigbear.com/bb2way
Bishop & Associates.	http://www.bishopinfo.com
The Bishop Co.	http://www.explainers.com

Black & Veatch	http://www.bvtelecomcom
Bosch Telecomn Inc.	http://www.boschtelecominc.com
Boston Com Group.	http://www.bcginc.com
Boston Technology.	http://www.bostontechnology.com
Boulder Blimp Co.	http://www.boulderblimp.com
Brand Comm. Ltd.	http://www.brandcomms.com
Brightpoint Inc.	http://www.brightpoint.com
Brite Voice Systems Inc.	http://www.brite.com
Broadband Networks Inc.	http://www.broadbandnetworks.com
Broadcast Buildings Inc.	http://www.broadcast-buildings.com
Broadcast Towers Inc.	http://www.btowers.com
Buena Vista TeleCom	http://www.utech.net/buenavista
Butler Telecom	http://www.butler.com
Butterfly Comm. Inc.	http://www.butterfly.com
Buttonwood Enterprises.	http://www.73.com/12v
C&D PowerCom	http://www.cdpowercom
C-Spec Corp	http://www.c-spec.com
Cable Prep.	http://www.cableprep.com
CableData Telecomm Inc.	http://www.laa.com
Cadence Design Systems.	http://www.cadence.com
Cadex Electronics.	http://www.cadex.com/cadex
CAL Corp.	http://www.calcorp.com
California Microwave Inc.	http://www.calmike.com
California Wireless Inc.	http://www.wireless.com
Call of the Wind.	http://www.windcall.com
Call One Inc.	http://www.call-1.com
Call Sciences.	http://www.callsciences.com
Call Technologies Inc.	http://www.calltec.com
Canamex Com Corp.	http://www.canamexcomcom
Canyon State Com	http://www.canyonstate.com
Capital Technologies.	http://www.capitaltechnologies.com
Capital Two-Way Com	http://www.capitaltwoway.com
Car Phone Express.	http://www.aawireless.com
Carolina Global Maps.	http://www.lbagroup.com/cgmaps
Carrier Comm..	http://www.av.qnet.com/~carcomm
Cartronix Inc.	http://www.cartronix.com
Cartwright Com Co.	http://www.cartwright.com
Cast-Tech.	http://www.case-tech.com
Castle Tower Corp.	http://www.castletower.com
Cbill Inc.	http://www.cbill.com
CCM Inc.	http://www.ccmcellular.com
CCS Int. Ltd.	http://www.spyzone.com
CDI Telecom Inc.	http://www.cdicorp.com
Celcore.	http://www.celcore.com
Celfone Paging Services Inc.	http://www.celfone.com
Cellcom	http://www.cellcomcom
Cell-Loc.	http://www.cell-loc.com
Cellmart.	http://www.cellmart.com
Cellmax.	http://www.rmscomm.com/cellmax
Cellnet Cellular.	http://www.cellnetcell.com
CellNet Data Systems Inc.	http://www.cellnet.com
Cellport Labs Inc.	http://www.cellport.com
Cellstar Corp.	http://www.cellstar.com
CellTalk Inc.	http://www.debitphone.com
Cell-Tel Int. Inc.	http://www.cell-tel.com
CellTrain.	http://www.celltrain.com
Cellular Accessory Warehouse	http://www.cawproducts.com

Cellular America	http://home.earthlink.net/~cellamdist
Cellular Doctor.	http://www.celldoc.com
Cellular Plus Inc.	http://www.wireless2yourdoor.com
Cellular Warehouse.	http://www.cellularwarehouse.com
Cellular Wholesalers Inc.	http://www.cellularwholesalers.com
Cellular Works.	http://www.cellworks.com
Cellxion Inc.	http://www.cellxion.com
Celplan Technologies Inc.	http://www.celplan.com
Celwave Inc.	http://www.celwave.com
Centerpointe Comm..	http://www.cencomcom
Centigram Com Corp.	http://www.centigram.com
Central Tower Inc.	http://www.centraltower.com
Century Telecom Inc.	http://www.cticallcenter.com
Ceotronics Inc.	http://www.ceotronics.com
Cerulean Technology Inc.	http://www.cerulean.com
CES Wireless Technologies.	http://www.ceswireless.com
Cetecom Global Service Net.	http://www.cetecomcom
Chadmoore Wireless Group	http://www.chadmoore.com
Cherokee Wireless Ser. Inc.	http://www.cherokeewireless.com
Chicago Comm. Service.	http://www.chicomm.com
Cibernet.	http://www.cibernet.com
Cirrus Logic Inc.	http://www.cirrus.com
CIS Com LLC.	http://www.ciscomm.com
CISS Ltd.	http://www.cissltd.com
Citel America Inc.	http://www.citelprotection.com
Citywide Enterprises Inc.	http://www.citywide.com
Cliffco Wireless.	http://www.cliffco.com
CLS Group.	http://www.capitalland.com
CMC Distributing Inc.	http://www.cmcmkt.com
CMC North.	http://www.cmcnorth.com
CML Technologies Inc.	http://www.cmltech.com
CNET Inc.	http://www.cnetinc.com
Coastcom	http://www.coastcomcom
Coherent Comm..	http://www.coherent.com
Columbia Capital Corp.	http://www.colcap.com
Comarco Wireless Tech. Inc.	http://www.comarco.com
Comm Site Int.	http://www.commsite.com
CommQuest Comm. Inc.	http://www.commquestinc.com
CommQuest Tech. Inc.	http://www.commquest.com
Commscope Gen. Inst.	http://www.commscope.com
CommSource Inc.	http://www.commsource.org.
CommStream Inc.	http://www.comstream.com
Comm. Inte. Corp.	http://www.cic.com
Comm. Resources Inc.	http://www.comresource.com
Comm. & Power Ind.	http://www.cpii.com/satcom/index.htm
Comm. Data Services Inc.	http://www.comm-data.com
Comm. Engineering Tech. Inc.	http://www.cetinc.com
Comm. Northwest, Inc.	http://www.commnw.com
Comm. Signaling Inc.	http://www.comsig.com
Comm. Systems Inc.	http://www.comm.systems.com
Comm. Test Design Inc.	http://www.ctdi.com
Comm. USA.	http://www.commusainc.com
CommuniTech.	http://www.communitech.com
Commware.	http://www.commware.com
Comp Comm Inc.	http://www.compcomminc.com
Compaq Computers Inc.	http://www.compaq.com
Compuquest Inc.	http://www.compuquestinc.com

Computer Innerworks.	http://www.voiceware.net.
Com-Rad Industries	http://members.aol.com/untenna
Comsat Corp.	http://www.comsat.com
Comsat Labs.	http://www.comsat.com/corp/lab
Comsat Personal Comm..	http://www.comsat.com/planet1
ComStream Corp.	http://www.comstream.com
ComTech Comm. Inc.	http://www.comtechcomm.com
Comtech Wireless.	http://www.comwireless.com
Comverse Technologies Inc.	http://www.comverse.com
Condor Comm. Inc.	http://www.condor-cci.com
Conductus Inc.	http://www.conductus.com
ConectUS Wireless.	http://www.conectus.com
Connect Systems Inc.	http://www.connectsystems.com
Contemporary Const. Inc.	http://www.ccitele.com
Convergence Corp.	http://www.convergencecorp.com
Conway Engineering Inc.	http://www.crl.com/~conway
Conxus Comm..	http://www.conxus.com
Corp Ten International.	http://www.corpten.com
Corsair Comm. Inc.	http://www.corsair.com
Counterpane Systems.	http://www.counterpane.com
Courtney Company Inc.	http://www.execpc.com/~dcpower
Crocomm Corp.	http://www.crocomm.com
Cromack Ind.	http://www.cromack.com
Cross Comm. Inc.	http://www.crosscomcom
Crown Comm..	http://www.crowncomm.net
Cruise Technologies.	http://www.cruisetech.com
Crystronics Inc.	http://www.pageco.com
CSI Shelter Technologies.	http://www.sheltech.com
CSS Power.	http://www.csspower.com
CTI Inc.	http://www2.tsixroads.com/~cti
CTI Information Services.	http://www.ctiis.com
CTP Systems Inc.	http://www.dspc.com
CTS Technical Services Co. Inc.	http://www.ctstech.com
Cushcraft Corp.	http://www.cushcraft.com
Custom Computer Services.	http://www.ccsinfo.com
CVDS Inc.	http://www.cvdsinc.com
Cyber Sites LLC.	http://www.cybersitesllc.com
Cybernet Paging.	http://www.cybernetpaging.com
Cylink Corp.	http://www.cylink.com
D&D Battery Inc.	http://www.4batteries.com
D&L Comm..	http://www.dlcomcom
Dannet	http://www.dannet.dk.
Danet Inc.	http://www.danet.com
DAPA.	http://www.dapacomcom
Data Critical Corp.	http://www.datacrit.com
Data Express.	http://www.dataexp.com
Data On Air Inc.	http://www.dataonair.com
Datamatic.	http://www.datamatic.com
Dataradio.	http://www.dataradio.com
Data Comm. Technologies.	http://www.dctcorp.com
Data-Tech Institute.	http://www.datatech.com
Datumtech Corp.	http://www.datumtechcorp.com
Dave Joyner & Associates.	http://www.davejoyner.com
Davicom Technologies.	http://www.davicomcom
Davis Electronics Co. Inc.	http://www.twoway.com
Day Management Corp.	http://www.wirelesscomm.com
DCI.	http://www.dcico.com

Decibel Products.	http://www.decibelproducts.com
DeftCom Inc.	http://www.deftcomcom
Delta Products Corp.	http://www.deltaca.com
Desktop Paging Software Inc.	http://www.desktoppaging.com
DGI Technologies Inc.	http://www.dgi-tech.com
Diablo Comm. Inc.	http://www.diablo.org.
DialAmerica Inc.	http://www.dialamerica.com
Dialogic Comm. Corp.	http://www.dccusa.com
Dialalogic Corp.	http://www.dialogic.com
Dialtone Discount Wireless.	http://www.dial-tone.com
Digital Microwave.	http://www.dmcwave.com
Digital Ocean Inc.	http://www.digitalocean.com
Digital Trans. Systems Inc.	http://www.dtsx.com
Digital Wireless Corp.	http://www.digitalwireless.com
Directc Power Plus.	http://www.dpp.com
Distributed Networks Inc.	http://www.dinetdata.com
Diva Comm..	http://www.diva.com
Diversified Comm. Tech.	http://www.flashtrack.com
Doppler Systems.	http://www.dopsys.com
DPS Inc.	http://www.dpstele.com
DSC Comm. INC.	http://www.dsccc.com
DSP Comm. Inc.	http://www.dspc.com
DTC Comm. Inc.	http://www.dtccomcom
DTS Wireless Inc.	http://www.dtswireless.com
D.W. Electrochemicals Ltd.	http://www.stabilant.com
Dynamic Instruments.	http://www.dynamicinst.com
Dynamic Mobile Data.	http://www.dmdsys.com
Dynacom TeleComm. Corp.	http://www.dyna-comcom
Dynamic TeleComm. Inc.	http://www.dynatele.com
Eagle Eye Tech. Inc.	http://www.eagle-eye-tech.com
Eastern Research Inc.	http://www.erinc.com
East View Cartographic.	http://www.cartographic.com
EcoSite Corp.	http://www.ecosite.com
EDX Eng. Inc.	http://www.edx.com
E.F. Johnson Co.	http://www.efjohnson.com
Electro Rack.	http://www.electrorack.com
Electro-Comm Distributing.	http://www.electro-com
ElectroCom Comm. Sys.	http://www.ecomsys.com
Electrodata Inc.	http://www.electrodata.com
Electronic Products Inc.	http://www.electpdcts.com
EmergiTech Inc.	http://www.emergitech.com
Emmitt Smith Comm.	http://www.emmittsmithcomcom
Empire Scientific.	http://www.empirebat.com
Empower Geo.	http://www.empowergeo.com
EMS Wireless.	http://www.emswireless.com
Enviro-Buildings Inc.	http://www.envirobldgs.com
Eon Corp.	http://www.eon.com
EPS Wireless.	http://www.eps-wireless.com
Ericsson Inc.	http://www.ericsson.com
Eupen Cable USA Inc.	http://www.eupen.com
Eventide.	http://www.eventide.com
Evolving Systems Inc.	http://www.evolving.com
Excel Inc.	http://www.xl.com
Expert Wireless Solutions.	http://www.expertwireless.com
Falcon Direct.	http://www.falcondirect.com
FAR Systems Inc.	http://www.farsystems.com
Federal Engineering Inc.	http://www.fedeng.com

Fibrebond Corp.	http://www.fibrebond.com
First Accessories.	http://www.first-acc.com
First Page.	http://www.firstpage.net
First Search Inc.	http://www.firstsearch.com
Flash Technology.	http://www.flashtechnology.com
Fox Electronics.	http://www.foxonline.com
FreeWave Technologies Inc.	http://www.freewave.com
French & Parrello.	http://www.fpawww.com
Fryer Site Gudes & Data Services.	http://www.fryersguide.com
Future Vision Inc.	http://www.futurevision.net
FWT Inc.	http://www.fwtinc.com
Federal Comm. Commission.	http://www.fcc.gov
FCC Wireless TeleComm. Bureau.	http://www.fcc.gov/wtb/siting.html
Gabriel Electronic, Inc.	http://www.gabrielnet.com
Gaia Software.	http://www.gaia.com
Galaxy PCS.	http://www.galaxypcs.com
Gambler-Johnson.	http://www.coredcs.com/~gamberj
Gandec Systems Inc.	http://www.gandec.com
Gateway Comm. Inc.	http://www.gateway.mexol.com
Geac Public Safety Inc.	http://www.geac.com
Gemplus.	http://www.gemplus.com
General Comm. Inc.	http://www.gencom
General Magic Inc.	http://www.genmagic.com
General Services Administration	http://www.gsa.gov/pbs/antenna.htm
The Genesis Group.	http://www.genesisworld.com
Gensym Corp.	http://www.gensym.com
Geodyssey Ltd.	http://www.geodyssey.com
GeoTrans Wireless.	http://www.geotranswireless.com
Geoworks.	http://www.geoworks.com
Glenayre Tech. Inc.	http://www.glenayre.com
GLM Wireless Comm.	http://www.glm-online.com
Global Access Pagers.	http://www.gapagers.com
GlobalComm Holdings.	http://www.globecom
Globalstar LP.	http://www.globalstar.com
GMSI Inc.	http://www.gmsisa.com
GNB Technologies.	http://www.gnb.com
GNP Computers.	http://www.gnp.com
GoAmerica Comm..	http://www.goamerica.net
GoComm Wireless Corp.	http://www.gocom
GoPage Inc.	http://www.gopage.net
GP Batteries USA.	http://www.gpbatteries.com
GPD Services Inc.	http://www.gpdco.com
Graychip Inc.	http://www.graychip.com
Grayhill Inc.	http://www.grayhill.com
Grayson Electronics.	http://www.grayson.com
GRF Inc.	http://www.gtcrf.com
Griffin Comm. Inc.	http://www.gcinetworks.com/wireless
GTE Intelligent Net. Svcs.	http://www.gteins.net
GTE Supply.	http://www.gtes.com
GTE Tele Svcs.	http://www.tsi.gte.còm
GTE Telephone Operations.	http://www.gte.com/carrier
GTE Wireless.	http://www.gte.com/c/chwire.html
GWCom Inc.	http://www.gwcomcom
Hamilton Electronics.	http://www.hamilton.net
Hammett & Edison Inc.	http://www.h-e.com
Hand Held Products Inc.	http://www.handheld.com
Harger Lightning Pro. Inc.	http://www.harger.com

Hark Systems Inc.	http://www.harksystems.com
Harris	http://www.farinon.harris.com
Harris Network Support Sys.	http://www.nss.harris.com
Hawker Energy Products Inc.	http://www.hepi.com
Hello Direct.	http://www.hello-direct.com
Henkels & McCoy Inc.	http://www.henkelsandmccoy.com
Hewlett Packard Co.	http://www.hp.com
Hitachi America Ltd.	http://www.hitachi.com
HiTech Wireless.	http://www.hitechwireless.com
Hittite Microwave Corp.	http://www.hittite.com
HM Electronics.	http://www.hme.com
HN Telecom Inc.	http://www.hntelecomcom
Hoag Electronics Inc.	http://members.aol.com/thoag
Holdsworth & Assoc. Inc.	http://www.holdsworth.com
Hughes Net Systems Inc.	http://www.hns.com
Hutton Comm. Canada.	http://www.huttcan.com
Hutton Comm. Inc.	http://www.huttoncomcom
Hybrid House Inc.	http://www.hybridhouse.com
The Hyde Co. Inc.	http://www.spatch.com
IBM Wireless Systems.	http://www.networking.ibm.com/wireless
ICOM America	http://www.icomamerica.com
IDA Corp.	http://www.idaco.com
IEX Corp.	http://www.iex.com
IFR Systems Inc.	http://www.ifrsys.com
Illuminet.	http://www.illuminetss7.com
Imperial Poles Div.	http://www.kellypipe.com
IMS Comm. Inc.	http://www.imsconnect.com
Industrial Two Way Radio, Inc.	http://www.twoway.net
InfoRad Inc.	http://www.inforad.com
Information Echo Inc.	http://www.infoecho.com
The Information Factory.	http://www.vllabs.com
Information Sto. Devices Inc.	http://www.isd.com
Information Tech. Inc.	http://www.itiusa.com
InformationView.	http://www.informationview.com
InfoSpace.	http://www.infospace.com
Infostream Pty Ltd.	http://www.infostream.comau
Infotech Solutions Corp.	http://www.infotechsol.com
Infotel International Inc.	http://www.infotel-intl.com
Innovations Inc.	http://www.innovationinc.com
Intasys Corp.	http://www.intasys.com
Integra Tech. Int. Inc.	http://www.integra.net
Integrated Site Dev.	http://www.isdc.net
Intek Global Corp.	http://www.intekglobal.com
Intel Corp.	http://www.intel.com
Intelect Net. Technologies.	http://www.intelectinc.com
Intelligent Information Inc.	http://www.intelligentinfo.com
IntelliNet Tech. Inc.	http://www.intellinet-tech.com
Intellisoft.	http://www.intellisoftonline.com
InterAir Wireless Inc.	http://www.interairwireless.com
InterDigital Comm. Corp.	http://www.interdigital.com
International Crystal	http://www.icmfg.com
International Marketing Sys.	http://www.ims-luke.com
International Mobile Comm.	http://www.worldcell.com
International Public Safety.	http://www.ingr.com/pubsafety
International Telecom Vent.	http://www.ejdort.com
International Tower.	http://www.internationaltower.com
InterPage.	http://www.iplp.com

InterServ TeleServices.	http://www.interservu.com
InterSky Comm.	http://www.intersky.com
InterState FiberNet.	http://www.itchold.com
Intertech Assoc. Inc.	http://www.intertechassociates.com
Intervoice Inc.	http://www.intervoice.com
In-Touch Mgt. Systems Inc.	http://www.itms.com
InTouch USA Inc.	http://www.intouchusa.com
Intraplex.	http://www.intraplex.com
Intuitive Software.	http://www.intuitive-software.com
Isis MPA.	http://www.inthetriangle.com/isis
IsoQuantic Tech. LLC.	http://www.isoquantic.com
ITS Corp.	http://www.itscorp.com
JJS Leasing Corp.	http://www.jjsleasing.com
JRC Canada Inc.	http://www.jrcc.com
JVC Co. of America.	http://www.jvc-america.com
K&L Microwave.	http://www.klmicrowave.com
Kasten Chase Applied Research.	http://www.kastenchase.com
Kaval Telecom Inc.	http://www.kaval.com
KCI Comm., Inc.	http://www.kcicomcom
Kenan Systems.	http://www.kenan.com
Kenwood Comm. Corp.	http://www.kenwood.net.
King Comm. USA.	http://www.kingusa.com
Klein Electronics.	http://www.kleinelectronics.com
Koast Comm. Corp.	http://www.koast.com
Kohler Co.	http://www.kohlerco.com
Kokusai Comm. Systems Inc.	http://www.kcswireless.com
KVH Industries Inc.	http://www.kvh.com
KVX Corp.	http://www.kvx.com
KW Microwave Corp.	http://www.kwmicrowave.com
Kyocera America Inc.	http://www.kyocera.com
La Marche Manufacturing Co.	http://www.lamarche-power.com
Lagoon Comm..	http://www.lagooncomcom
Lama Electronics Inc.	http://www.lamaelectronics.com
Landmark Tower Corp.	http://www.landmarktower.com
Larsen Electronics Inc.	http://www.larsenet.com
Larus Corp.	http://www.laruscorp.com
LBA Group Inc.	http://www.lbagroup.com
LBA Technology Inc.	http://www.lbagroup.com
Lease Corp. of America.	http://www.leasecorp.com
LeBlanc. Inc.	http://www.leblanctelecomcom
Lenberg Comm. Inc.	http://www.lenberg.com
Let's Talk Cellular & Wireless.	http://www.letstalk.com
Lett Electronics.	http://www.lettelectronics.com
LGP Inc.	http://www.lgp-inc.com
LHS Comm. Systems Inc.	http://www.lhsgroup.com
LHS Group Inc.	http://www.lhsgroup.com
Liberty WireStar Inc.	http://www.libertywire.com
Lightbridge Inc.	http://www.lightbridge.com
Lightning Eliminators & Cons. Inc.	http://www.lightningeliminators.com
Linderlake Corp.	http://www.linderlake.com
Linktronic Systems Inc.	http://www.linktronics.com
Litton Solid State.	http://www.littonsolidstate.com
Locsys Inc.	http://www.locsys.com
Locweld Inc.	http://www.locweld.ca
Lodestar Towers Inc.	http://www.lodestartowers.com
Logica Inc.	http://www.logica.com
Logus.	http://www.logusmicrowave.com

Lorch Microwave.	http://www.lorch.com
Lucent Tech. Inc.	http://www.lucent.com
LXE Inc.	http://www.lxe.com
M&C Systems Inc.	http://www.mcsys.com
MaComCo.	http://www.macomco.com
Magellan Systems.	http://www.magellangps.com
Malouf Engineering International.	http://www.mei1inc.com
Manchester International Inc.	http://www.miiweb.com
Manon Engineering Inc.	http://www.manon.com
Marconi Instruments Inc.	http://www.marconi-instruments.com
Mark/Space Softworks.	http://www.markspace.com
Marketronics Corp.	http://www.marketronics.com
Mason Electric Co.	http://www.mason-electric.com
Matthews Land Co. Inc.	http://www.matthewsland.com
Maxon America Inc.	http://www.maxonusa.com
Maxrad Inc.	http://www.maxrad.com
MCComm..	http://www.mccomcom
McKenzie TeleComm. Group.	http://www.mtg.com
McManus Comm..	http://www.mcmanuscomm.com
McQuerter Group.	http://www.mcqtr.com
Megatec International.	http://www.megatecusa.com
Melard Technologies Inc.	http://www.melard.com
Mentor Engineering Inc.	http://www.mentoreng.com
Meridian Geographics.	http://www.mgmap.com
Meridian Wireless Tech. Inc.	http://www.meridianwireless.com
MET Laboratories Inc.	http://www.metlabs.com
Metapath Software Corp.	http://www.metapath.com
Metawave Comm..	http://www.metawave.com
Metric Equipment Sales.	http://www.metricsales.com
Metric Systems Corp.	http://www.metricsystems.com
Metrica Inc.	http://www.metrica.com
Metricom Inc.	http://www.metricomcom
Metro One TeleComm. Inc.	http://www.metro1.com
MFJ Enterprises Inc.	http://www.mfjenterprises.com
MicroAge, Mobility Solutions.	http://www.microage.com/mobility
Microcell Management Inc.	http://www.mcmsite.com
Micronetics Wireless.	http://www.mwireless.com
Microphase Corp.	http://www.microphase.com
Microspace Comm.	http://www.microspace.com
Microware Systems Corp.	http://www.microware.com
Microwave Data Systems.	http://www.mdsroc.com
Microwave Filter Co. Inc.	http://www.microwavefilter.com
Microwave Solutions.	http://www.mwsolns.com
Midamerica Tower Leasing Inc.	http://www.towern.com
Miller Building Systems Inc.	http://www.millerbldgs.com
Millitech Corp.	http://www.millitech.com
Mitsubishi Elect. America Inc.	http://www.melamsat.com
Mitsubishi Wireless Comm. Inc.	http://www.mitsubishiwireless.com
Mizco Int. Cellular Innovations.	http://www.mizco.com
MJA Comm. Corp.	http://www.mjacommcorp.com
MLJ Inc.	http://www.mlj.com
MMA Inc.	http://www.mmaconsult.com
Mobile 2-Way Service Inc.	http://www.flex.net/u/m2w.
Mobile Comm.	http://www.mobilecomm.com
Mobile Data Initiative.	http://www.gsmdata.com
Mobile International	http://www.mobileintl.com
Mobile Mark Inc.	http://www.mobilemark.com

The Mobile Phone Co.	http://www.mobilephonecompany.com
Mobile Solutions Inc.	http://www.mobisol.com
Mobile Telestructures.	http://www.mobilemini.com
Mobilecomm	http://www.mobilecomm.com
MobileLogic Network Services.	http://www.mobilelogic.com
MobileQuest Wireless.	http://www.mobilequest.com
MobileWare Corp.	http://www.mobileware.com
Modems Plus Inc.	http://www.modemsplus.com
Modular Technology Systems.	http://www.modtechsys.com
Momentum Telecom Inc.	http://www.momentumtele.com
Moseley Associates Inc.	http://www.moseleysb.com
Motorola Inc.	http://www.mot.com
MTU Wireless Accessories Inc.	http://www.mtuwireless.com
Murandi Comm..	http://www.murandi.com
MX-Com Inc.	http://www.mxcomcom
Nady Systems.	http://www.nadywireless.com
Narda Microwave.	http://www.nardamicrowave.com
Natcom Inc.	http://www.natcominc.com
National Safety Equipment.	http://www.safety-source.com
National Semiconductor.	http://www.national.com
National TeleComm. & Info.	http://www.ntia.doc.gov
Nationwide Tower Co.	http://www.watertank.com
Natural MicroSystems.	http://www.nmss.com
Navtech GPS Supply.	http://www.navtechgps.com
NDC Voice.	http://www.ncdwireless.com
NEC America Inc.	http://www.nec.com
Nemal Elec. Int. Inc.	http://www.nemal.com
New Media Communication.	http://www.nmcfast.com
New World Power Technology Co.	http://www.nwvt.com
Newmar.	http://www.newmarpower.com
Newnet Inc.	http://www.newnet.com
NEX Comm.	http://home.att.net/~nexcomm
Nextek Inc.	http://www.ultranet.com
The NLS Group.	http://www.nlsgroup.com
Noise Cancellation Tech. Inc.	http://www.nct-active.com
Noise Com Inc.	http://www.noisecomcom
Nokia Corp.	http://www.nokia.com
North America TeleComm. Corp.	http://www.natctelecomcom
Northern Telecom Inc.	http://www.nortel.com
Northpoint Comm. Products Inc.	http://www.nadtech.com/northpoint
Not Limited Inc.	http://notlimited.com
NotePage Inc.	http://www.notepage.com
Nova Engineering.	http://www.nova-eng.com
Novatel Wireless Inc	http://www.novatelwireless.com
NSI.	http://www.nsiradio.com
NSN Network Services	http://www.nsn.net.
Nutel Electronics Corp.	http://www.hoflink.com/~nutel
Oak Frequency Control Group.	http://www.ofc.com
Oki Semiconductor.	http://www.okisemi.com
Okitele Comm.	http://www.oki.co.jp.
Oldcastle Precast.	http://www.oldcastle-precast.com
Omega Digital Data Inc.	http://www.omegadigital.com
OmniBrowse Inc.	http://www.omnibrowse.com
Omnipoint Corp.	http://www.omnipoint.com
Omnitrend.	http://www.omnitrend.com
On Target Mapping	http://www.otmapping.com
Open Development Corp.	http://www.opendev.com

OpenLink Corp.	http://www.open-link.com
Optaphone Systems.	http://www.optaphone.com
Option International.	http://www.option.com
Opto Electronics.	http://www.optoelectronics.com
Opus Telecom Inc.	http://www.opustelecomcom
Ora Electronics.	http://www.orausa.com
Oracle Corp.	http://www.oracle.com
Orbital Sciences Corp.	http://www.orbital.com
Orga Card Sys.	http://www.orga.com
Ortel Corp.	http://www.ortel.com
OTG Cellular.	http://www.otgcellular.com
Otto Comm..	http://www.ottoeng.com
P-com M Inc.	http://www.p-comcom
Pacific Access.	http://www.pacacc.com
Pacific Comm. Sciences Inc.	http://www.pcsi.com
Pacific Crest Corp.	http://www.paccrst.com
Pacific Monolithics Inc.	http://www.pacmono.com
Pacific Moon Trading Co.	http://www.silcomcom/pacmoon.
Pacific West.	http://www.adm.net/~pwe
Pactronics.	http://www.pactronics.com
PageCo Int. Inc.	http://www.pageco.com
Pagecorp Industries.	http://www.pagecorp.com
PageNet.	http://www.pagenet.com
Pager Automation.	http://www.automation.net
Pager Parts Unlimited Inc.	http://www.pageco.com
PageRepair Inc.	http://www.pagerep.com
Pagers Plus Cellular.	http://www.pagers.com
Pagers Unlimited.	http://www.pagersunlimited.com
PageStar Inc.	http://www.pagestar.com
PageTap Inc.	http://www.pagetap.com
PAL Tele Comm.	http://www.palgroup.com
Paragon Networks Int.	http://www.paragon-networks.com
Paragon Software Ltd.	http://www.paragonsoftware.com
Parallel Comm.	http://www.parallel-comm.com
Parwan Electronics Corp.	http://www.voicesaver.com
PC Guardian.	http://www.pcguardian.com
PC Outlet.	http://www.pcoutlet.net.
PC Paging Inc.	http://www.pcpaging.com
PCCard Packaging Inc.	http://www.pccardpkg.com
PCS Technologies.	http://www.pcs-tech.com
PCS Telecom Inc.	http://www.pcstelecomcom
PCS Wireless Inc.	http://www.pcswire.com
Peiker Acustic Inc.	http://www.peiker.com
Penncomm USA Inc.	http://www.penncommusa.com
Percomm Pagers.	http://www.percomm.com
Periphonics Corp.	http://www.peri.com
Persoft Inc.	http://www.persoft.com
Personal Comm. Inc.	http://www.pagepci.com
Phoenix Wireless Group Inc.	http://www.pwgi.com
Photocomm Inc.	http://www.photocomm.com
Piezo Crystal Co.	http://www.piezo-crystal.com
Pirod Inc.	http://www.pirod.com
Pocket News.	http://www.pocketnews.com
Point to Point Comm. Inc.	http://www.ppcila.com
Polyphaser Corp.	http://www.polyphaser.com
Positron Inc.	http://www.positronpower.com
The Pouch Inc.	http://www.sisna.com/users/pouch.

Power & Telephone Supply.	http://www.ptsupply.com
Power Conversion Products Inc.	http://www.pcpinc.com
Power Inc.	http://www.power-inc.com
Power-Sonic Corp.	http://www.power-sonic.com
Powerwave Tech. Inc.	http://www.powerwave.com
Prairie Systems.	http://www.prairiesys.com
Precision Systems Inc.	http://www.talkpsi.com
PrePaid Wireless Services Inc.	http://www.prepaidcell.com
Price Comm. Inc.	http://www.pricecomm.com
Priority Call Mgt.	http://www.prioritycall.com
Prism Telecom.	http://www.tga.com
Procomm.	http://www.proComm..com
Pro-Fit International Inc.	http://www.pro-fit-intl.com
Professional Comm..	http://www.bfrog.com/pcs
Professional Pride Training Co.	http://www.911train.com
Professional Software of Amarillo.	http://www.psasoft.com
Professional Tech. Sys.	http://www.ptsystems.com
Profile Marketing Research.	http://www.profile-mktg-res.com
Profile Paging.	http://www.profilepaging.com
Progressive Concepts Inc.	http://www.hawkexpress.com
Project Group 2000.	http://www.projectgroup2000.com
Protocol TeleComm. Inc.	http://www.radiolink.com
Proxim Inc.	http://www.proxim.com
PSI International.	http://www.psiint.com
Psion Software PLC.	http://www.software.psion.com
PT Systems.	http://www.ptsystems.com
Public Safety Associates Inc.	http://www.psadallas.com
Public Safety Systems Inc.	http://www.publicsafetyinc.com
Public Safety Wireless Network.	http://www.pswn.gov
Pulse Power.	http://www.pulsepower.com
Q-Safety Inc.	http://www.qsafety.com
Q.E.D. Inc.	http://www.qed.com
Qualcomm Inc.	http://www.qualcomm.com
Quest Systems Inc.	http://www.questsystems.com
Quinsy BV.	http://www.quinsy.com
Qum Software.	http://www.qum.com
Ra-Tel Comm., Inc.	http://www.buyratel.com
Rabun Labs Inc.	http://www.rabunlabs.com
Racal Antennas.	http://www.racal.com
Racal Instruments Group.	http://www.racalinst.com
Racom Corp.	http://www.racomnet
Racon Inc.	http://www.racon.com
Racotek Inc.	http://www.racotek.com
Radio Comm. Wholesalers.	http://www.radiocomm.com
Radio Design Group.	http://www.radiodesign.com
Radio Dynamics Corp.	http://www.radyn.com
RadioFone.	http://www.radiofone.com
RadioLAN.	http://www.radiolan.com
RadioMail Corp.	http://www.radiomail.net
Radionics, Inc.	http://www.radionic.com
Raine Inc.	http://www.raineinc.com
Raltron Electronics Corp.	http://www.raltron.com
Ramsey Electronics Inc.	http://www.ramseyelectronics.com
RAM Technologies.	http://www.ramlink.net.
Ratelco Electronics Inc.	http://www.ratelco.com
Raytheon Wireless Solutions.	http://www.raytheon.com/micro
RCS Comm. Group.	http://www.rcscomcom

Readycom Inc.	http://www.readycomcom
Realogic.	http://www.realogic.com
Real Time Strategies Inc.	http://www.rts-inc.com/rts
ReCellular Inc.	http://www.recellular.com
Record/Play Tek Inc.	http://www.cyberlink-inc.com/rpt
Recoton Corp.	http://www.recoton.com
Redcom Labs Inc.	http://www.redcomcom
Reflection Technology.	http://www.reflection.com
Reltec Corp.	http://www.relteccorp.com
Repeater Technologies.	http://www.repeaters.com
Reproduction Technologies Inc.	http://www.cyberlink-inc.com/rti
Research in Motion.	http://www.rim.net.
RF Imaging & Comm..	http://www.best.com/~rfimage
RF Marketing.	http://www.rfmarketing.com
RF Micro Devices Inc.	http://www.rfmd.com
R.F. Tech. Pty. Ltd.	http://www.rftechnology.comau/rfinfo
R.G. Sound & Comm..	http://www.rgsound.com
Richardson Electronics Ltd.	http://www.rell.com
The Right Co.	http://www.rightcompany.com
Riser Bond Instruments.	http://www.riserbond.com
Ritron Inc.	http://www.ritron.com
Rittal Corp.	http://www.rittal-corp.com
RMS Comm. Group Inc.	http://www.rmscomm.com
Rockwell Int. Corp.	http://www.cacd.rockwell.com
Rohn.	http://www.rohnnet.com
R.S. Comm.	http://www.rs-comm.com
Russet Wireless Group.	http://www.usa1.com/russet.
Ryno Enterprises.	http://www.futureway.com
Safco Tech. Inc.	http://www.safco.com
Sage Inst. Inc.	http://www.testservices.sageinst.com
Salisbury Engineering.	http://www.salisbury-engineering.com
Sanders.	http://www.sanders.com
SaRonix.	http://www.saronix.com
Saville	http://www.savillesys.com
Scala Electronic Corp.	http://www.scala.net
SCC Comm. Corp.	http://www.scc911.com
Scientific-Atlanta Inc.	http://www.sciatl.com
Scoden Comm..	http://www.scoden.com
S.C.G. Ltd.	http://www.peachking.com
Scopus.	http://www.scopus.com
SCT.	http://www.reachsct.com
SEA Inc.	http://www.sea-dmi.com
Search4Sites.	http://www.search4sites.com
Seavey Engineering Assoc.	http://www.seaveyantenna.com
Secure Dev. Network Inc.	http://www.smsmail.com
Seiko Comm. of America, Inc.	http://www.messagewatch.com
Selectone Inc.	http://www.selectone.com
Sentry Manufacturing Co.	http://www.sentrymfg.com
Serviceware Corp.	http://www.serviceware.ca.
SGC U.S. Testing Co.	http://www.ustesting.sgsna.com
Shaffer Comm. Group Inc.	http://www.shafcomm.com
Shively Labs.	http://www.shively.com
Shure Brothers.	http://www.shure.com
SicolaMartin.	http://www.sicolamartin.com
Siemens.	http://www.siemens.com
Siemens Solar Industries.	http://www.solarpv.com
Sierra Wireless Inc.	http://www.sierrawireless.com

Signal Group.	http://members.aol.com/signalgrp
Signal Intelligence.	http://www.scanstar.com
Signal One.	http://www.signal-one.com
Signal TeleComm. Ins. Services.	http://www.thesignal.com
Signal Tower Comm., Inc.	http://www.signaltower.com
Signius Corp.	http://www.signius.com
Silverlake Comm. Inc.	http://www.silverlake2000.com
Sima Products Corp.	http://www.simacorp.com
Sinclair Tech Ltd.	http://www.sinctech.com
Sinvan Co.	http://www.sinvanco.com
Sioux Falls Tower Spec. Inc.	http://www.anchorguard.com
Site-Patcher.	http://members.aol.com/sitepatch
SiteMaster.	http://www.sitemaster.com
Skylar Wireless Comm..	http://www.skylar.org.
SLA Distributors Inc.	http://www.slawireless.com
Slattery Software.	http://www.ewol.com/slattery
SmarTrunk Systems Inc.	http://www.smartrunk.com
Smith Myers USA Inc.	http://www.smithmyers.com
SMR Direct.	http://www.smrdirect.com
SNC Mfg. Co. Inc.	http://www.sncmfg.com
Snider Tele.	http://www.snider.net
Socket Comm. Inc.	http://www.socketcomcom
SoftWright LLC.	http://www.softwright.com
Solar Depot Inc.	http://www.solardepot.com
Solar Electric Specialties.	http://www.solarelectric.com
Solectek Corp.	http://www.solectek.com
SoloPoint Inc.	http://www.solopoint.com
Sony Electronics Corp.	http://www.sony.com
Sound Solutions Inc.	http://www.clubmaid.com
Source Search.	http://www.sourcesearch.com
Southwest Microwave Inc.	http://www.smicomcom
Spacecom Systems.	http://www.spacecomcom
Sparton Technology Inc.	http://www.spartontech.com
SpectraSite Comm..	http://www.spectrasite.com
Spectrian Inc.	http://www.spectrian.com
Spectrum Resources Towers LP.	http://www.spectrumresources.com
Spillman Data Systems Inc.	http://www.spillman.com
Sprint North Supply	http://www.sprintnorthsupply.com
Spyglass Inc.	http://www.spyglass.com
SQL Systems.	http://www.dstm.com
SSI Services Inc.	http://www.vanadium.com
Standard Telecom.	http://www.nixxopagers.com
Stanford TeleComm. Inc.	http://www.stelhq.com
StatPower Technologies Corp.	http://www.statpower.com
Stay Tuned.	http://www.staytuned.com
Stealth Net. Tech. Inc.	http://www.stealthsite.com
Strategic Open Solutions Group.	http://www.sos.com
Strongwell.	http://www.strongwell.com
Summa Four Inc.	http://www.summa4.com
Summit Manufacturing Inc.	http://www.summitmfginc.com
Sun Manufacturing Inc.	http://www.sunmfg.com
Sun Microsystems.	http://www.sun.com
Sunbelt TeleComm. Inc.	http://www.1800sunbelt.com
SunGard Mailing Service.	http://www.sungard.com/sms.
Super Sonic Systems.	http://pages.prodigy.com/supersonic.
Super Wireless Warehouse.	http://www.exposhow.com/wireless/541.htm
Supra Products Inc.	http://www.keysafe.com

Survey Tech. Inc.	http://www.surveytech.com
Sutter Buttes 2-Way.	http://www.2-way.com
Swift Computers Inc.	http://www.swiftcomp.com
Switzer Products Corp.	http://www.switzerproducts.com
SWS Security.	http://www.swssec.com
Syclo Corp.	http://www.syclo.com
Symbol Technologies Inc.	http://www.symbol.com
SymmetriCom Inc.	http://www.symmetricomcom
Synacom Technology Inc.	http://www.synacomcom
Synergy Microwave Corp.	http://www.synergymwave.com
Synetcom Digital Inc.	http://www.synetcomcom
System Perfect Inc.	http://www.sysperfect.com
T-Netix Inc.	http://www.t-netix.com
Taconic Adv. Dielectric Div.	http://www.taconic-add.com
Tait Electronics-USA Inc.	http://www.taitus.com
Talley Comm. Corp.	http://www.talleycomcom
Tascomm Inc.	http://www.tascomm.com
TCC Industries Inc.	http://www.tccinc.com
TCSI Corp.	http://www.tcsi.com
TEA Group Inc.	http://www.teagroup.com
TEC Cellular Inc.	http://www.tecc.com
TechnaSource.	http://www.technasource.com
Tech-Ni-Comm Inc.	http://www.cellular-phone.com
Technical Equip. Distrib. Inc.	http://radiodata.com/tedi.
TechnoCom Corp.	http://www.technocom-wireless.com
Technology Trends.	http://www.technology-trends.com
Technostores.	http://www.technostores.com
Tecore Inc.	http://www.tecore.com
TekNow Inc.	http://www.teknow.com
Tektronix Inc.	http://www.tek.com
Telamon Corp.	http://www.telamon.com
Telco Systems Inc.	http://www.telco.com
Telecom Analysis Systems Inc.	http://www.taskit.com
Telecom Solutions.	http://www.telecomcom
Trlrcom Towers Inc.	http://www.telecomtowers.com
TeleComm. Sys.	http://www.tcsnet.net.
TeleComm. Eng. Assoc.	http://www.tcomeng.com
TeleComm. Power Sys. Inc.	http://www.tpspwr.com
TeleComm. Techniques Corp.	http://www.ttc.com
Telect Inc.	http://www.telect.com
Teledesign Systems Inc.	http://www.teledesignsystems.com
Telemac Cellular Corp.	http://www.telemac.com
Telemobile Inc.	http://www.telemobile.com
TelePath Corp.	http://www.tele-path.com
Telesource Inc.	http://www.telesourcenet.com
TeleStructures Inc.	http://www.telestructures.com
Tele-Tech Co. Inc.	http://www.tele-tech.com
Teletronics Int. Inc.	http://www.teletronics.com
Telewave Inc.	http://www.telewaveinc.com
Teleworx.	http://www.teleworx.com
Telex Comm..	http://www.telex.com
Tellabs Inc.	http://www.tellabs.com
Telogy.	http://www.tecentral.com
Telogy Networks Inc.	http://www.telogy.com
Telrad.	http://www.telrad.com
Teltrust Inc.	http://www.teltrust-inc.com

Telstra Corp.	http://www.telstra.com
Telular Corp.	http://www.telular.com
Telxon Corp.	http://www.telxon.com
Ten-Four Comm..	http://www.ten-four.com
Terra Comm. Inc.	http://www.terracomm.com
Tessco Tech. Inc.	http://www.tessco.com
Test-Tronics.	http://www.refurb.com
Texas Comm.	http://www.texascomcom
Texas Inst.	http://www.ti.com
Texas Microsystems.	http://www.texmicro.com
Thompson Tower Ins. Inc.	http://www.thompsontower.com
Thunder Eagle Inc.	http://www.thuneagle.com
Tiburon Inc.	http://www.tibinc.com
Times MW Sys.	http://www.timesmicrowave.com
TMA Data Inc.	http://www.tmadata.com
TMC Comm..	http://www.tmc4radios.com
Tocad America Inc.	http://www.tocad.com
Toll Free Cellular.	http://www.tollfree.com
Tomba Comm..	http://www.tomba.com
Total Software Sol. Inc.	http://www.totalsoftware.com
Totally Wireless.	http://www.totallywireless.com
Tower Resources Mgt. Inc.	http://www.trmcom.com
Tower Structures Inc.	http://www.towerstructures.com
Tower Systems Inc.	http://www.towersystems.com
Trachte Channelframe Bldg. Inc.	http://www.trachte-channelframe.com
Trackit Software.	http://www.trackit.com
TRAK Microwave Corp.	http://www.trak.com
Trans Pac Batteries.	http://www.transpacbatteries.com
Trans-Tech.	http://www.alphaind.com
TranSettlements Net. Svs. Inc.	http://www.transettlements.com
Traverse Bay Environ.	http://www.tbgroup.com/tbe.html.
TruePosition.	http://www.trueposition.com
TrueTime Inc.	http://www.truetime.com
Trylon-TSF.	http://www.trylon.com
TSR Paging Inc.	http://www.beep.com
TTC.	http://www.ttc.com
Tunnel Radio of America.	http://www.tunnelradio.com
Tusonix Inc.	http://www.tusonix.com
TuWay Wireless.	http://www.tuway.com
TWR Lighting Inc.	http://www.twrcompanies.com
TWR Telecom Inc.	http://www.twrcompanies.com
Tycor Internatinoal Corp.	http://www.tycor-intl.com
U.S. Dept. of Commerce.	http://www.ita.doc.gov/industry/tai/telecom/telecom.html
U.S. Information Agency	http://www.usia.gov/usis.html
U.S. Sec. & Exchange Comm.	http://www.sec.gov
Uniden America Corp.	http://www.uniden.com
UniDial Comm..	http://www.unidial.com
Unisys.	http://www.unisys.com
United Telecom Inc.	http://www.imi.on.ca
Universal Shielding.	http://www.nell.com
Unplugged Comm.	http://www.upcomm.com
UNR-Rohn.	http://www.rohnnet.com
Unwired Data Inc.	http://www.unwireddata.com
Unwired Planet.	http://www.uplanet.com
UP Inc.	http://www.upinc.com

U.S. Robotics Inc.	http://www.usrobotics.com
U.S. Sitelease Inc.	http://www.us-sitelease.com
USA Mobile Comm. Inc.	http://www.usamobile.com
USCS International.	http://www.uscs.com
Utilicom Inc.	http://www.utilicomcom
Valcom Inc.	http://www.valcominc.com
Valmont Microflect.	http://www.comtowers-poles.com
Vari-L Co.	http://www.vari-l.com
Varta Batteries.	http://www.varta.com
Vega Signaling Prod.	http://www.eviaudio.com
Ventera Corp.	http://www.ventera.com
Versatility Inc.	http://www.versatility.com
ViaSat Inc.	http://www.viasat.com
Vicorp Inter. Systems.	http://www.vicorp.com
VLSI Technology Inc.	http://www.vlsi.com
VoCom Products Co.	http://www.vocomrf.com
Voice Processing Corp.	http://www.vpro.com
Voiceware.	http://www.voiceware.net
Wacom Products Inc.	http://www.wacomprod.com
Walker and Associates Inc.	http://www.walkerassoc.com
Watkins-Johnson Group	http://www.wj.com
Wavetek Corp.	http://www.wavetek.com
WCM Electronics.	http://wcmelectronics.com
West Coast Eng. Ltd.	http://www.wceng.com
Westec Comm. Inc.	http://www.westeccomcom
Westinghouse Wireless.	http://www.wireless.westinghouse.com
Whisper Comm. Inc.	http://www.whisper.com
Wi-LAN Inc.	http://www.wi-lan.com
Windata Inc.	http://www.windata.com
Wink Comm..	http://www.wink.com
Winstar.	http://www.winstar.com
Wireless Access.	http://www.waccess.com
Wireless Inc.	http://www.wireless-inc.com
Wireless InterNet Corp.	http://www.twin.net.
Wireless Internet LLC.	http://www.wirelessisp.net
Wireless North.	http://www.wirelessnorth.com
Wireless Services Corp.	http://www.airnote.net.
Wireless Tech. Group.	http://www.pageco.com
Wizard Software.	http://www.commware.com
W&W Associates.	http://www.wwassociates.com
Wynd Comm. Corp.	http://www.wynd.com
Xcel Controls Inc.	http://www.xcel.com
Xemond Inc.	http://www.xemond.com
Xertex Technologies Inc.	http://www.xertex.com
Xetron Corp.	http://www.xetron.com
Xircom Inc.	http://www.xircomcom
XL Computing Corp.	http://www.xlcomputing.com
Xtend Comm. Corp.	http://www.xtend.com
Yaesu/Vertex.	http://www.yaesu.com
Yuasa-Exide Inc.	http://www.yuasa-exide.com
Zero Enclosures.	http://www.zerocorp.com
Zetyron Inc.	http://www.zetron.com
ZK Celltest Inc.	http://www.zkcell.com

APPENDIX C GLOSSARY OF TERMS

This Glossary defines many of the terms expressed in italics used in this *Guide*. It is based upon industry standard, vendor and common industrial definitions and terms published frequently in trade journals and advertisements. Consult the *Wireless Week* Web page for additional terms. **www.wirelessweek.com**

A

A-Band cellular- Cellular system assigned to non-wireline.
A/B Switch- A feature found on newer cellular telephones permitting the user to select either the "A" (non-wireline) carrier or the "B" (wireline) carrier when roaming away from home.
Access fee- Fee to connect with the local phone network.
Adaptive antenna- A smart antenna that shapes the pattern of a wireless system.
AIN- Advanced Intelligent Network a system for multimedia messaging, PCS and cell routing.
Air interface- Wireless network operating system i.e. AMPS, TDMA, CDMA and GSM.
Airtime- Actual time used on a wireless system.
A-key- Subscriber authentication number.
Alphanumeric- A text readout.
AMPS- Advanced Mobile Phone Service the standard for analog cellular service.
Analog- Method of modulating radio signals to carry information. Most common are AM (amplitude modulation) and FM (frequency modulation).
ANSI- The American National Standards Institute, a non-profit, privately funded membership organization that coordinates the development of standards.
Antenna- A device for radiating and/or receiving RF signals.
Antenna Farm - Group of towers in a near location.
ATM- Asynchronous Transfer Mode a high speed & bandwidth transmission method.
Auto PC- An in-vehicle system using Windows CE from Microsoft Corp.

B

B-Band cellular- Cellular system assigned to the telco operating subsidiary.
B-Block Carrier:-A 30 Mhz PCS carrier serving a Major Trading Area (MTA) in the frequency range 1870-1885 MHz. paired with 1950-1965 MHz.
Base Station- The central radio transmitter/receiver that maintains communications with mobile radiotelephones within a given range (typically a cell site).
Bandwidth- A relative range of frequencies that can carry a signal without distortion on a transmission medium.
B-CDMA- Broadband code division multiple access a technology developed by InterDigital Comm. Corp.
Bent pipe technology- Satellite method to transmit calls from one point on Earth to a satellite and back down to another point.
Bidding credits- Discount sometimes given to small businesses in FCC spectrum auctions.
Boomer Site- High profile high power base station.
Broadband- A wide-bandwidth channel for voice, data and/or video services.
Broadband PCS- Personal communications services in the A- through F-Block auctions and used for voice and data.
BTA- Basic Trading Area composed of several contiguous counties. BTAs are service areas designed by Rand McNally and adopted by the FCC. There are 493 BTAs in the United States.
Bundling- Combining many telecom services into one simplified package.

C

CAD- Computer Aided Dispatch commonly used by public-safety agencies. Also, Computer Aided Design.
Caller ID- A feature that displays a caller's phone number on the wireless handset.
Calling party pays- Call originator is billed for call to a wireless phone.
Cap code- Electronic identification number used to signal a pager.
C-Block Carrier- A 30 Mhz PCS carrier serving a Basic Trading Area in the frequency range 1895-1910 MHz. paired with 1975-1990 MHz.
Cell Splitting- A means of increasing the capacity of a wireless system by subdividing or splitting cells into two or more smaller cells.
CGSA- Cellular Geographic Service Area is

the actual area in which a cellular company provides cellular service. This area may be smaller than the MSA or RSA surrounding it.
CDMA- Code division multiple access is a spread spectrum system used in some digital cellular and personal communications services.
CMDA One- A standard, coined by the CDMA Development Group.
CDPD- Cellular Digital Packet Data An overlay for transmitting and receiving data over a cellular system.
Cell site- The location where the wireless antenna and communications equipment is placed.
Chip set- A combination of two or more integrated circuits on one module.
Churn- A measure of the number of subscribers who leave to go to another carrier's service.
ClassLink- A program of the CTIA Foundation providing wireless phones to schools .
CLEC- Competitive Local Exchange Carrier provides local wireline phone service.
Clone- A wireless phone programmed with stolen or duplicated electronic serial and mobile identification numbers.
CMRS- Commercial mobile radio service is the FCC designation for any carrier or licensee whose wireless network is connected to the public switched telephone network.
Co-branding- The use of two or more different brand names on a single product.
Co-location- Placement of multiple antennas at a common physical site to reduce environmental impact and real estate costs.
Cost Recovery- Reimbursement to CMRS providers of both recurring and nonrecurring costs.
"Covered" SMR: A subset of specialized mobile radio operators subject to a particular set of regulations.
COW- Cell site on wheels is a mobile site placed at a location to fill in or increase coverage.
CPE- Consumer premise equipment are telephones, PBXs and other communications devices located in the home or office.
CTIA- Cellular Telecommunications Industry Association is a trade group representing cellular, PCS and enhanced specialized mobile radio carriers.
Customer acquisition cost- The average cost to a carrier of signing up an individual subscriber.

D

D-AMPS- or digital AMPS used by Ericsson Inc. to describe IS-136 time division multiple access technology.
D-Block Carrier- A 10 MHz. PCS carrier serving a Basic Trading Area in the frequency range 1865-1970 MHz. paired with 1945-1950 MHz.
DCS 1800- A digital cellular system for PCS mobile communications network used outside of the U.S.
DCMA- Dynamic channel multi carrier architecture is a technology developed by ComSpace Corp.
De minimus- Having such minimal impact as to not influence the matter at hand.
Digital Modulation- A method of encoding information for transmission. Information is turned into a series of digital bits. Digital transmission offers a cleaner signal and is virtually immune to the problems of analog modulation such as fading and static.
Digital signal processor- A microprocessor that digitizes analog signals.
Disaggregation- The splitting of an FCC license into two or more licenses of fewer frequencies.
Drive test- A method of sampling signal strength measurements in a mobile coverage area.
DTV- Digital television is the next generation of video and audio technology for TV broadcasters.
Dual band- Describes a handset that works on 800 MHz cellular and 1900 MHz PCS frequencies.
Dual mode- Describes a handset that works on both analog and digital networks.

E

E911- Enhanced 911 is a 911 service with automatic number identification and automatic location information provided to the 911 operator.
EA license- Economic area license defined for 176 "economic areas" delineated by the Bureau of Economic Analysis at the U.S. Department of Commerce.
E-Block Carrier- A 10 MHz. PCS carrier serving a Basic Trading Area in the frequency range 1885-1990 MHz . paired with 1965-1970 MHz.
ECO test- Effective competitive opportunities test developed to determine whether U.S. carriers enjoy effective competition in a foreign carrier's market.

EDACS- Enhanced digital access communications system is a private radio or specialized mobile radio network designed by Ericsson Inc.

Electronic serial number- The unique identification number embedded in a wireless phone by the manufacturer.

Encryption- The process of "scrambling" a message to prevent it from being read by unauthorized parties.

ETACS- Extended Total Access Communications Systems is the conventional wireless technology used in the United Kingdom and other countries. It was developed from the U.S. AMPS technology.

ERMES- European radio messaging system is a paging system used in Europe.

ESMR- Enhanced specialized mobile radio is a digital SMR network, usually referring to Nextel Communications Inc. and Geotek Communications Inc., that provide dispatch, voice, messaging and data services.

ETSI- European Telecommunications Standards Institute.

Ex parte- Statements, meetings or filings that are made *outside* of an official comment/or reply period.

F

F-Block Carrier- A 10 MHz. PCS carrier serving a Basic Trading Area in the frequency range 1890-1995 MHz paired with 1970-1975 MHz.

FDMA- Frequency Division Multiple Access is a method of radio transmission that allows multiple users to access a group of radio frequency bands without interference.

FCC- Federal Communications Commission is the federal agency responsible for commercial and private spectrum management.

FHMA- Frequency hopping multiple access is a digital technology used in Geotek Communications Inc.'s ESMR network.

FLEX- A Motorola Inc.-licensed protocol that provides carriers more capacity on their networks and faster transmission times. Also refers to the FLEX family of protocols: FLEX, InFLEXion and ReFLEX.

Flash memory- A component used for memory that can retain information without power.

FNPRM- Further notice of proposed rule making is a document issued by the FCC to solicit additional comments.

Frequency- A measure of the energy, as one or more waves per second, in an electrical or light-wave information signal. A signal's frequency is stated in either cycles-per-second or Hertz (Hz).

G

Gateway- Ground-based link to a mobile satellite.

Geosynchronous- A fixed orbit, about 24,000 miles above the Earth.

GPS- Global positioning system is a series of 24 geosynchronous satellites that continuously transmit their position.

GSM- Global system for mobile communications is a digital cellular or PCS network used throughout the world.

GSM-Plus- An enhanced version GSM to meet IMT-2000 capabilities.

GHz.- GigaHertz is a billions Hertz. Personal Communications Services operate in the 1.9 GHz band of the electromagnetic spectrum.

Guyed- A type of wireless transmission tower that is supported by cables.

3G - 2Ghz. multimedia network transmitting wireless data at 144 kps at mobile user speeds, 384 kbps at pedestrian user speeds and 2 megabits per second in fixed locations.

H

Handoff- Switching a mobile call to an adjacent cell site.

HDML- Handheld device markup language allows Internet access from wireless devices.

HLR- Home location register a database residing in a local wireless network that checks the identity of a local subscriber.

Hertz- A measurement of electromagnetic energy, equivalent to one "wave"or cycle per second.

I

IEEE- The Institute of Electrical and Electronics Engineers who establishes standards for wireless local area network interoperability.

iDEN- Integrated digital enhanced network is a Motorola Inc. ESMR network technology that combines two-way radio, telephone, text messaging, and data transmission into one system.

ILEC- Incumbent local exchange carrier is the local phone service provider in a market.

Immunity- Not being criminally liable or liable for any damages in a civil action for injuries,

death or loss to person or property resulting from any act or omission.
IMT-2000- An ITU name for a third generation global standard for mobile telecommunications.
IMTA- International Mobile Telecommunications Association is the trade group serving SMR and PAMR carriers.
InFLEXion- Is the narrowband PCS system developed by Motorola that allows for voice paging.
Interconnection- The connecting of one network with another.
Interoperability- The ability of a network to operate with other networks.
IS- Interim standard is the designation of the American National Standards Institute that refers to an accepted industry protocol.
IS-41- The network standard that allows all switches to exchange information about subscribers.
IS-95- The standard for code division multiple access.
IS-54- The first generation of the digital standard time division multiple access technology.
IS-136- The latest generation of the digital standard time division multiple access technology.
ISDN- Integrated services digital network is a high-capacity wireline data system.
ITA- Industrial Telecommunications Association is a trade group serving private wireless licensees.
ITU- International Telecommunication Union is the agency of the United Nations that furthers the development of telecommunications.
IXC- Interexchange Carrier is a long-distance phone company.

K

KHz- Kilohertz is one thousand Hertz. Each wireless phone call occupies only a few KiloHertz.

L

LAN- Local area network is a client computer connected to a server.
LEC- Local exchange carrier is a wireline phone company.
LEO- Low earth orbit for a mobile communications satellite.
LifePage- A service of the PCIA Foundation that donates pagers to individuals awaiting for organ donor transplants.
Little LEO- A low-earth orbiting satellite system primarily providing data services.
LMCC- Land Mobile Communications Council is a trade group of frequency coordinators and associations serving private users and commercial operators.
LMDS- Local multipoint distribution service is a broadband radio service designed to provide two-way transmission of voice, high-speed data and video.
Local calling area- Is the local no toll calling area.
Local number portability- The ability of subscribers to switch local or wireless carriers and still retain the same phone number.

M

Messaging- Text paging, e-mail or short messages received on alphanumeric wireless devices.
Microcellular- Process of directing the signals to isolated areas.
MIPS- Millions of instructions per second describes digital signal processing.
MIN- Mobile identification number is a unique mobile number not the same as an ESN.
MTSO- Mobile telephone switching office.
Moratoriums- Waiting periods on the issuance of construction permits by local zoning authorities.
MOU- Minutes Of Use directly affecting revenue. Also Memorandum of Understanding.
MSA- Metropolitan statistical area is the coverage area of a city for a given cellular network.
MHz.- MegaHertz is one million Hertz Cellular and ESMR systems operate in the 800 and 900 MHz. bands of the electromagnetic spectrum.
Monopole- A slender self-supporting tower on which wireless antennas can be placed.
MSC- Mobile Switching Center also MTSO.
MTA- Major trading areas (51 in the US) are composed of several contiguous basic trading areas (BTA).
Mutually Exclusive- Two or more applications for the same spectrum use rights.

N

NAM- Number assignment module is the electronic memory in the wireless phone that stores the telephone number and electronic serial number.
NATE- National Association of Tower

Erectors.

NAMPS- Narrowband advanced mobile phone system combines cellular voice processing with digital signaling, to increase capacity.

Narrowband PCS- Is another generation of paging network, including two-way, acknowledgment and talk back paging.

Notice of Inquiry- The first step in an FCC rule making.

NEPA- The National Environmental Policy Act of 1969 is an Act of Congress which requires Federal Agencies to take into consideration the potential environmental effects of a particular proposal such as construction of a radio station.

Nondirectional Antenna- An antenna that transmits and receives equally well in all directions, usually on one plane; also called an omnidirectional antenna.

Non Ionizing Radiation- Radio waves, infrared rays and visible light rays, none of which can affect an atom's electrical balance.

Non-wireline Cellular Company- The Block "A" carrier. The "A" originally stood for "alternate" system. The FCC, in setting up the licensing and regulatory rules for cellular, decided to license two cellular systems in each market. It reserved one for the local telephone company and opened a second system -- the Block A system -- to other interested applicants. The distinction between Block A and Block B was meaningful only during the licensing phase at the FCC. Once a system is constructed, it can be sold to anyone. Thus in some markets today, both the A and B systems are owned by telephone companies.

NTIA- National Telecommunications and Information Administration is the federal government's spectrum management authority.

Number pooling- A method for maximizing the use of phone numbers.

Numeric- A display, message or readout that contains numbers usually associated with pagers..

O

Off-peak- Portion of the day air time rates might be lowered.

Operating Margin- A measurement of a company's relative profitability calculated by dividing operating profit by net sales.

P

Partitioning- Parceling a spectrum license into two or more geographic areas.

Pay phone service provider- A company that owns and operates pay phones.

PC Card- Formerly called a PCMCIA card, a detachable card that can be plugged into a laptop or personal computer.

PCIA- Personal Communications Industry Association is the trade group representing PCS, SMR, private radio and other wireless users and carriers. See also SOMA.

PCS- Personal communications services are two-way, 1900 MHz digital offerings.

PDA- Personal digital assistant is a portable computing device capable of transmitting data.

Peak- That part of the day that where mobile phone rates might be higher.

Picocell- A cellular base station with extremely low output power designed to cover an extremely small area, such as one floor of an office building.

POCSAG- Refers to an open standard developed by the U.K.'s Post Office Code Standards Advisory Group and widely used in this country for numeric paging.

POPs - Persons of population describing the number of potential customers.

Preemption- When a Federal agency voids a local ordinance or state law, asserting that the Federal government, not the state or locality, has jurisdiction in the matter.

Prepaid cellular- Subscribers pay in advanced for wireless service.

Price/earnings ratio: A measurement of a company's value as expressed by the stock market. P/E is a company's stock price divided by its net income per share.

Project 25- Is a joint government/industry effort to set standards for the development of technical standards for another generation of two-way communications equipment.

Project Angel- AT&T Corp.'s code name for a wireless local loop residential network.

Proportional and owned subscribers- Some wireless carriers share markets in partnership with others so they report their subscribers as a percentage.

PSAP- Public-safety answering point such as an E 911 dispatch center.

PSTN- Public switched telephone network is the worldwide voice telephone system.

PSWAC- Public Safety Wireless Public Utility Commission is the name for the state regulatory body charged with regulating utilities including telecommunications.

R

Radio-frequency fingerprinting- A process that identifies a mobile phone by its unique transmission switch print.
RBOC- Regional Bell operating company.
Reciprocal billing- Wireline companies pay for their cost of terminating originated calls.
Recon petition- Petition for reconsideration of an FCC decision.
Refarming - An FCC initiative to promote more efficient use of the frequency bands below 512 MHz, allocated to private land mobile radio services.
ReFLEX 25- A narrowband PCS technology developed by Motorola allowing two-way text messaging.
ReFLEX 50- A technology that denotes the proprietary two-way system implemented by Mobile Telecommunication Technologies Inc.
Repeater- Device that receives a radio signal, amplifies and re-transmits it in a new direction.
Return on equity- An indication of how much a company is earning on the investment its shareholders have made in it. Divide net income by shareholder equity. Also ROI or return on investment.
RF- Radio Frequency is a frequency well above the range of human hearing.
Roaming- The ability to use a wireless phone to make and receive calls in places outside one's home calling area.
RSA- Rural service area is the non-metropolitan area covered by a cellular licensee.

S

Satellite phone- A wireless phone that uses a mobile satellite.
Shareholder rights provision or "poison pill"- A stock plan intended to discourage or prevent an unwanted, "hostile" corporate takeover.
Shielding- The use of a particular material to shield an electronic component or circuit from the effects of external electric or magnetic fields.
SIM- Subscriber identity module or smart card.
Simulcast- A paging technique that broadcasts the same signal over each site in a network at the same time.
Smart antenna- An antenna system whose technology enables it to select a desired signal.
Smart card- A plastic card containing important data about a person's identity permitting access.
SMR- Specialized mobile radio is a dispatch radio and interconnect service for businesses.
SMS- Short message service.
Soft key- A key below the phone's display that performs whatever function is listed on the display.
SOHO- Small office/home office defines a market that consists of individuals that work out of their homes or other offices.
SOMA- Site operators and managers association, a subgroup of PCIA.
Spectrum allocation- Federal government designation for a block of frequencies used in a category of service.
Spectrum assignment- Federal government authorization for use of specific frequencies or frequency pairs within a given allocation.
Spread spectrum- a transmission technology which spreads information over a much greater bandwidth than would be necessary.
SRAM- Static random access memory is a memory technology used in pagers and handsets.
Standby time- The amount of time a subscriber can leave a fully charged handset turned on to receive incoming calls before the batteries go dead..
Subordinated debt- A lien that is considered secondary to a senior obligation.
Subscriber fraud- An illegal deception to secure wireless service with intent to avoid payment.
Subscriber profiling: Is the process of of compiling subscriber usage information in order to identify potentially fraudulent use.

T

TCP/IP- Transmission control protocol/Internet protocol is the Internet protocol suite which governs the exchange of sequential data.
TDMA- Time division multiple access is a digital air interface technology used in cellular, PCS and ESMR networks.
Telecommunications Act of 1996- Signed into law on February 8, 1996, which establishes and deregulates a pro-competitive framework for telecommunications in the United States.
Termination charges- Fees that wireless telephone companies pay to complete calls on wireline phone networks or vice versa.
TETRA- Terrestrial trunked radio is an open digital trunked radio standard defined by the

European Telecommunications Standardization Institute.

Thin client- A pen-based tablet computer used on a wireless local area network.

Third-generation- A new standard that promises to offer increased capacity and high-speed data applications up to 2 megabits.

Tri-mode handset- Phones that work on three frequencies, typically using 900 MHz, 800 MHz digital or reverting to 800 MHz. analog cellular when digital is not available.

Triple band- A network infrastructure or wireless phone designed to operate in three frequency bands.

Turnkey- An entire system with hardware and software assembled and installed by a vendor and sold as a total package.

U

Un-validated phone- A wireless phone lacking a service contract or which is unregistered.

UWCC- Universal Wireless Communications Consortium is an industry group supporting IS-136 time division multiple access and IS-41 wireless intelligent network technology.

V

VLR- Visitor location register is a database that holds information about roaming customers.

Voice activation- A feature that allows a subscriber to dial a phone by spoken commands.

Vocoder- A device used to convert speech into digital signals.

Voice Mail- A computerized answering service that answers a call, plays a greeting and records a message.

VSAT- Very small aperture terminal is a small satellite dish installed at end-user locations.

W

W-CDMA- Wide band code division multiple access has been proposed for a new generation of CDMA .

WCS- Wireless communications services in the 2.3 GHz band and designated for general fixed wireless use.

WIN- Wireless intelligent network which provides enhanced and customized services for mobile telephones.

Windows CE- Software platform for handheld personal computers, developed by Microsoft Corp.

Wireless- Using the radio-frequency spectrum for transmitting and receiving voice, data and video signals for communications.

Wireless Internet- An RF-based service that provides access Internet e-mail and/or the World Wide Web.

Wireless LAN- A local area network which uses a radio-frequency connection for an office-based computer network.

Wireless PBX- Equipment that allows employees or customers within a building or limited area to use wireless handsets connected to an office's private branch exchange system.

WLL- Wireless local loop is a fixed service that competes with or substitutes for local wireline phone service.

World Radio Conference: Biennial meetings of the ITU member nations to discuss and resolve global spectrum allocation issues.

Z

Zero coupon bond: A debt instrument that pays all accrued interest at the bond's maturity instead of making regular cash payments to the bondholder during the life of the bond.

APPENDIX D   ASSOCIATIONS & PUBLICATIONS

American Mobile Telecommunications Association.
http://www.amtausa.org.
Association for Local Telecommunications Services.
http://www.alts.org.
Association of Public-Safety Communications Officials International.
http://www.apcointl.org.
CDMA Development Group.
http://www.cdg.org.
CDPD Forum.
http://www.cdpd.org.
CTIA.
http://www.wow-com.com.
CompactFlash Association.
http://www.compactflash.org.
Consumer Electronics Manufacturers Association.
http://www.cemacity.org.
Electronics Industry Association.
http://www.eia.org.
Forest Industries Telecommunications.
http://www.landmobile.com.
Foundation for Rural Service.
http://www.frs.org.
Global Wireless Education Association.
http://mnic.net/~gwec.
GSM Association.
http://www.gsmworld.com.
Industrial Telecommunications Association.
http://www.ita-relay.com.
Institute of Electrical and Electronics Engineers Inc.
http://www.ieee.org.
International Mobile Telecommunications Association.
http://www.imta.org.
National Association of Tower Erectors.
http://www.daknet.com/nate.
PCIA.
http://www.pcia.com.
Portable Computer and Communications Association.
http://www.outlook.com/pcca.
Satellite Industry Association.
http://www.sia.org.
Telecommunications Industry Association.
http://www.tiaonline.org.
Universal Wireless Communications.
http://www.uwcc.org.
UTC
http://www.utc.org.
Wireless Cable Association.
http://www.wirelesscabl.com.
Wireless Communications Alliance.
http://www.wca.org.

Wireless Dealers Association.
http://wirelessdealers.com.
Wireless Industry Association.
http://wirelessindustry.com.
The Wireless LAN Alliance.
http://www.wlana.com.

Publications

It is strongly recommended that you subscribe to these two publications immediately:

Wireless Week
P.O. Box 10554
Riverton, NJ 08076-8554
(303) 393-7449

RCR
Circulation Department
965 E. Jefferson
Detroit, MI 48207-9834
(800) 678-9595

Wireless Product News
P.O. Box 5360
Pittsfield, MA 01203-9788
(413) 637-4343
Wireless Week BUYER'S GUIDE
P.O. Box 10554
Riverton, NJ 08076-8554
(303) 393-7449
America's Network
Advanstar Communications Inc.
P.O. Box 6100
Duluth, MN 55806-9800
(218) 723-9437
Telecommunications
685 Canton Street
Norwood, MA 02062
(617) 762-9230
Radio Resource Magazine
PO Box 11774
Riverton, NJ 08076-9374
(609) 786-4415
Wireless Business & Technology
Phillips Business Information, Inc.
P.O. Box 3208
Northbrook, IL 60065-9889
(301) 309-3847
Phillips PCS SOURCE BOOK
1201 Seven Locks Road
Suite 300
Potomac, MD 20854
(301) 340-1520

Wireless World
P.O. Box 12928
Overland Park, KS 66282-2928
(913) 967-1903
Mobile Radio Technology
P.O. Box 12968
Overland Park, KS 66282-2968
(913) 967-1903
Newaves (in Personal Communications)
434 W. Downer Place
Aurora, IL 60506-9692
(312) 563-9919

Communications Products
1100 Superior Ave..
Cleveland, OH 44197-8137
Intrafamily Communications Inc.
P.O. Box 6093
Freehold, NJ 07728
Tower Times
420 4th. Street NE Suite 109
Watertown, SD 57201-2699

Newsletters:

All Phillips Publications:
PCS Week
Land Mobile Radio News
Wireless Data News
Mobile Phone News
www.TelecomWeb.com